VISIONS OF A BETTER WORLD

THE FUTURE IN ST[...]

*Brice
Le Blévennec*

Lannoo

T0093589

SUMMARY

By Brice Le Blévennec
Founder and CVO of Emakina Group

INTRODUCTION
DREAM BIGGER

'As inventors we're obliged to dream, to be unconstrained in our quest for progress. Always to be pushing at the barriers'.
Clive Sinclair, in the TV movie *Micro Men*

It started in 1980. I was a shy and introverted 13-year-old boy who had just entered secondary school at the Athenaeum Adolphe Max. My new school was one of three lucky athenaeums chosen for a pilot e-learning programme. One day the prefect, Jean Berger, announced the installation of Control Data's PLATO (Programmed Logic for Automatic Teaching Operations) terminals in a dedicated room. As I had managed to avoid the Dutch language classes, I started to spend all those free hours tinkering with this system: terminals connected to the world by rudimentary modems, long before the birth of the Internet as we know it today. The terminals displayed very precise vectors instead of large pixels. The screen was even touch-sensitive thanks to tiny wires that formed a grid of one centimetre squares. It was a platform that was 20 years ahead of its time.

DISCOVERING INCREDIBLE POTENTIAL

My first hack was a little programme in TUTOR that mimicked the first screen, where users entered their login and password to start the terminal. When I was leaving the room, I would launch my program that displayed a fake screen with the mention 'PRESS "NEXT" TO BEGIN' to trick future users. The next day, I would collect the passwords of everyone – including teachers – that had used the terminals in the interim. One day, I even collected the password of a user from group S: the engineers who maintained the system, and then I could start playing network games that were blocked at school. I played Moria, an adventure game in a rudimentary isometric 3D maze, with monsters and treasures. I discovered «Dogfight», the first 3D air combat simulator at 0.1

4

fps. I was already playing on a network with hundreds of geeks across Europe and the US. The system's experience was very basic, but my imagination was running wild. I anticipated 3D images as realistic as the ones I saw during the CGI sessions at the Cartoon Festival. I imagined a much faster network and I felt that there was the potential to change the world.

LOVE AT FIRST SIGHT

In 1982, during a school trip to London, I spent all my pocket money for food to buy a Sinclair ZX81. As soon as I got back, I connected it to an old TV set in the attic and started to leaf through the manual which was dedicated to BASIC programming. After devouring it from cover to cover, I fell into coding. It was love at first sight. I spent my nights coding Breakout or Pong, copying pages from magazines, until my father cut the fuses on my floor to force me to go to sleep.

'The most exciting phrase to hear in science,
the one that heralds new discoveries,
is not "Eureka!" but "That's funny..."'.

Isaac Asimov

After trying to code a version of the Moria game on a ZX Spectrum, I devoted myself to coding a Power 4, whose main quality was the multi-coloured interface and sound effects. The intelligence of the software was so rudimentary that it was impossible to lose against it. But I understood that the user interface design wowed players and could be used as an illusion to hide the flaws of a rather simple programme. I then devoured a book about the popularisation of computer languages, published in French by Eyrolles. I discovered Fortran, Pascal, Prolog and LISP. Then I got my first PC assembled at Infoboard, with my first database management tools – Dbase, Foxpro – and I sold my first software (an armory management system).

'With great power comes great responsibility'.

The Peter Parker principle,
in the *Spider-Man* comic books written by Stan Lee

I understood that computers gave humans superpowers and that I could use them to compensate for my weaknesses. I also understood that this revolution would profoundly change the world.

THE FRUIT OF KNOWLEDGE

In 1989, my mother, who ran a word processing office, bought a Macintosh SE/30 which I promptly monopolised. I fell in love with its windows, its menus, and the mouse which made the software intuitive. I used it to make magazines with SuperPaint. I was getting incredible results at the time. I used to go to our Apple dealer to print my work on a LaserWriter.

The shopkeeper was often impressed by my tinkering. One day in September, when I had just graduated – late – from high school, he offered me a job. The mission was to help a company that had just bought a brand-new, beautiful Macintosh IIfx for a few days. A total dream for me. I immediately accepted, and that's how I started a career as a graphic designer at Paparazzi, a below-the-line communications agency, where I worked for 18 months. I very quickly proved to be indispensable and converted the whole agency to Mac. While working at the agency, I had a short spell studying Typography and Graphic Design at La Cambre, a Brussels school of visual arts. Thereafter I moved on to computerising a photoengraving company with Scitex systems, which I connected to an Apple 'Tops' network to reduce costs and increase possibilities with the first versions of Photoshop. I learned image processing, engraving, offset – and how to face sleepless nights.

One day, as I was stopping by the Paparazzi agency to collect payment on a late bill, the head of the studio, Catherine Decarpentrie, offered me a job to create our little prepress office. I accepted, and that same night I chose its name: Ex Machina. After a period in a shared office, we established our first private office in a garage in Forest, a commune of Brussels. The first 10 years were crazy; we did pre-press, CD-ROMS, interactive terminals, then the first websites, and basic video editing. I was living on-site, next to my servers that I babysat at night. I was working non-stop, and I loved it.

FROM EX MACHINA TO EMAKINA

Over time, our cutting-edge technological innovation attracted prestigious clients – Belgacom, Coca-Cola, Electrabel, Apple, Swatch, to name but a few – and proposals to buy out our little agency came pouring in. But I was not ready to give up my independence and freedom. In 2001, a group of entrepreneurs who had founded an e-business agency proposed that we join forces to found Emakina. This became a reality 20 years ago on April 1st, 2001. Since that day, we have never stopped growing and in 2006 we went public. The confidence of the market gave us the means to undertake an international adventure, first in Europe, then in Asia, the USA and Africa. We are now present in 20 countries, on four continents.

I have always tried to use technological innovations in a creative way to generate value for clients. The world is a big place with many people brighter than me, so I was unlikely to invent anything that didn't already exist somewhere. But by exploiting the latest technological innovations in a very creative and original way, with clear objectives, strategy and a good plan, it's possible to create unique new services, applications or content that create value for a customer, ex nihilo.

'We believe in progress, scientific, rational.
And we believe that it must be at the service of humanity
and that man must never be at the service of technological
progress that has escaped him'.

Emmanuel Macron during his 2021 speech about *Plan France 2030*

Emakina believes in progress through putting strategy, technology and creativity at the service of users. This has been Emakina's corporate culture since day one. It's this culture that underpins everything we do – our investments, our recruitment of talent, and our discourse within the market. Emakina is The User Agency. If our projects are innovative and creative, they are adopted by users, and that makes our clients successful. If our clients are satisfied, they continue their partnership with us. It's a virtuous circle.

APPLIED SCIENCE FICTION

'We need dreamers and idealists, people who have an incredible and difficult vision of how the future fits together, to make things happen'.

Michael Dell about Steve Jobs

To stay ahead of the curve, we must look at users; we must anticipate their needs and wants to be ready for our clients. We must constantly imagine a positive future to build the path that leads to it. We chose to start with a user experience in the future in the form of a fiction. Then we came back to the present, to its scientific publications, its innovations, its trends – in the form of an essay. We found our North Star and then we built the path to go in its direction. It's a new genre that combines fiction and essay. I've called it 'Applied Science Fiction'.

Ex Machina was founded 30 years ago. So naturally, we asked ourselves where we would be in 30 years. This book explores what might happen between 2021 and 2051, in 30 articles devoted to 30 areas. Each article begins with a short fiction. In our agency language: a User Experience consisting of a situation lived by one imaginary character (or more). Then we share our vision of a possible future. In our language, we start with insights to establish our foresight. Finally, we look at the technological trends, and the recent innovations that make these stories more or less plausible. With your smartphone scan the QR code featured on the first page of each article to access full online references.

THREE APPROACHES TO THE FUTURE

Three Emakina teams have been working on this project, each with a specific area of expertise.

BLUE ARTICLES. Our content specialists have an investigative journalistic approach. They investigate the state of the art, the latest technological trends and scientific publications to come up with their stories. At the helm are Manon Dubreuil, Paula Fitzhenry, Jean-Christophe Detrain and Cédric Godart.

GREEN ARTICLES. Our User Experience consultants are all part of the DXD (Digital eXperience Design) team at Emakina.BE. Their stories are intimately linked to their understanding of future user needs. They take into account the sociological evolution of society, as well as the impact of technology to elaborate their scenario. The tandem consists of Content Designer Sarah Claeys, and UX researcher Iva Filipovic. Another DXD team-mate, Design Strategist Vicky De Mesmaecker, helped realise the 'RoboCop 2.0' article with her background and knowledge of criminology.

ORANGE ARTICLES. Our visionaries start from their imaginations, fed by their insatiable curiosity towards all sorts of subjects, and a technological intuition to discover improbable but often possible futures. This is the spitting image of Brice Le Blévennec and his talented sparring partner, Johannie van As.

FROM SCIENCE TO FICTION

The articles are ordered from the most probable to the craziest. Each article is placed on a scale from Science to Fiction.

| SCIENCE | INNOVATION | DISRUPTION | VISION | FICTION |

SCIENCE. The first articles are fairly close to the state of science and, barring accidents, their advent is highly probable.

INNOVATION. The following articles anticipate innovations that are fairly logical and, with the current acceleration in the pace of innovation, are fairly likely.

DISRUPTION. At the heart of the book, these articles explore possibilities that depend on radical innovations that are still at research stage.

VISION. These articles are predictions, disconnected from the feasibility of available technologies, but human inventiveness has no limits other than those of physics.

FICTION. The latest articles take the concept to its extreme. We will imagine the wildest possibilities in a world where today's physical barriers have been broken by scientific discoveries yet to come.

ART
& ALAN TURING

Maria is bored. As an entertainment lawyer, she has sat through some tedious negotiation meetings in her life, but this one takes the cake. It's only a few hours until she can escape to her absolute favourite past-time: attending an outdoor punk opera. Featuring AI punk band, The Last Eardrum, this sold-out show also includes a drone symphony designed by AI choreographer, Robot Wars. But the most exciting part? Debbie Harry will posthumously 'sing' a brand-new track, mashed up from her vast catalogue of songs. She will be accompanied by the PhilharSonic Orchestra, consisting of robot percussionists and humans on the strings. The track 'Maria' is playing softly in the background to get her in the mood. Maria wonders if there will be one million drone lights tonight... She traded one of her favourite digital art pieces for the concert ticket, so this had better be good!

But back to the task at hand. Maria's boss asked her to be physically present in this meeting as there's a lot of money at stake. 'Present', meaning she needs to show proof that she's locked out of the metaverse for that hour. 'Present' also by iris scan authentication, verifying that she is there (in the flesh) and it's not her digital twin. Yet, everyone else in the meeting is not human. On her screen, she is watching four legal bots argue about the artistic rights of her clients: the rock band AK40 Winks. She's finding it very boring because in her mind the case is cut and dried. Jon, Pol, Jorge and Ringoo are all established virtual entertainers who have decided to create this band for this two-album deal. Each bot is arguing for royalties for each of their clients, but Maria knows that the neural net that has written, composed and mixed this album is owned by the AK40 Winks brand. Therefore, all rights too. End of conversation.

As she listens to them drone on about legal precedents, the nano-sensor in her arm picks up that her heart rate and breathing are dropping (out of tedium, no doubt) so her personalised soundscape subtly accelerates in pace to lift Maria into action. Without noticing, she starts tapping her feet, then gets up and moves around the room, getting her blood circulation moving. Music mission accomplished.

A few hours later, when Maria has finally had enough, she tells the bots that AK40 Winks' proprietary neural network owns all the rights to these songs and that the solo artists will only be compensated for their performances. Now she can finally click 'end' and get ready for a night out. From the window of her apartment, a mesmerising sight catches her eye. The learning centre across the road comes alive at night with randomly selected poetry, photography, theatre and science experiments transformed into exquisite ever-evolving artworks.

This reminds her to check in on her son, Seb. Tapping her AR glasses into his view, she can see him working on his latest AI choreography project. He has been briefed to turn movements and poses of every Swan Lake performance ever filmed into an AI sequence the Botshoi (robot) ballerinas can perform. It's their first rehearsal and the robots are quite comically struggling with the Black Swan movement.

POWER TO THE MACHINES

Imagine if you could collaborate with AI on almost anything? In 2020, OpenAI – the Elon Musk-founded AI research lab – announced that the newest version of its AI system could mimic human language. The largest neural network ever created, GPT-3, is trained on (almost) the entire Internet and has revolutionised the AI world. It has opened up a world of opportunities. The results were astounding. Soon it was writing blogs, poetry, songs, and scripts. It was musing about the meaning of life. It was generating pick-up lines, creating dating profiles, and building apps.

And it wasn't trained to do any of these things. GPT-3 learned how to learn. It is 100x bigger than its predecessor, with a whopping 175 billion parameters. When it comes out, GPT-4 will not just be more powerful, it might be capable of true reasoning and understanding. It will probably handle a much larger context. We'll be able to feed it with video, audio, books – you name it. Now can you imagine the possibilities of GPT-20? All we know is that it will stake its claim firmly in the world of art.

By 2050, the art we consume will be spectacularly personalised and co-created by talented AI musicians and artists. Humans will share artistic authorship and recognition with machines. AI will blur the very definition of art. Music will be much more than entertainment but a proven form of precision medicine. Generative music apps will pull your biological, situational, brainwave and mood data to create a custom-made, ever-adapting playlist just for you. Only, it won't be music as you know it, but all-new soundscapes. With the flutter of our eyelids or the warmth of our breath, the soundscapes around us will adjust to keep us energised and comfortable. Songs won't be one-size-fits-all. Forward-thinking artists will realise that music can become a treatment and a 3D experience. Albums will be available as functional music – for example, to improve your memory, sharpen your focus or break bad habits. Paintings and sculptures will still be around, but artists will find the lure of the blockchain too irresistible not to dabble in digital art. Algorithms will be their new favourite pigment. Musicians will have hundreds of new sounds and instruments to play with – or simply use their bodies to create beats. You don't need to have a good voice or play an instrument to be a musician. You just need a vision. Lyrics and songs can be automatically generated from just a snippet of melody. AI will put together entire film scores, write soundtracks and set music to adverts.

Imagine how, in the world of marketing, you could tailor advertising soundtracks to a person's mood or location. Imagine being able to tweak your dinner party playlist as if you have hundreds of live bands right there in the room – jamming together. It's going to be such fun, we can't wait.

ARTIFICIAL CREATIVITY?

'Where computers complement the human creative process is that they create a type of beauty that is hard for humans to make on their own.' Spend some time on the Artbreeder and you'll see why its founder's words ring true. Inspired by evolution, artist Joel Simon's site uses a machine learning method known as a generative adversarial network (GAN) to manipulate portraits and landscapes. You can 'crossbreed' pictures of everything from animals and people to landscapes and objects to create uncanny and sometimes beautiful artworks. As if by magic, your portraits can be adjusted through simple sliders to change age, race, emotions and much more. Before you dismiss this as just another fun face filter app, think again. This is the same tool that helped designer Daniel Voshart bring ancient Roman emperors back to life, transforming statues into photorealistic faces – and getting academics all excited about this new data. GAN art is huge. Creations are not only visually stunning but earn their creators big bucks too. For instance, the artist collective Obvious fed an algorithm 15,000 images of portraits from different time periods. It then generated its own portraits (with some human intervention, of course), one of which sold for $432,500 at a Christie's auction. The algorithm is composed of two parts – a Generator that makes a new image based on the set, and a Discriminator which tries to spot the difference between a human and an AI-made image.

It's both forger and art detective. When the 'judge' is fooled, that's when the artists deem it a result, which was how the piece (*Portrait of Edmond Belamy*) was chosen.

This is an interesting notion, where the machine is both creator and critic. Another more impressive take on GAN is the 'creative' generative network built by the Art and Artificial Intelligence Lab at Rutgers University. AICAN is the first and only patented algorithm for making art using AI. It's trained on 100,000 of the greatest works in art history and it is specifically programmed to produce novelty – not emulate creativity like Obvious. From its data set of paintings from the 14th century, every creation is something completely different. The art is also largely abstract – as if it knows the direction of art's trajectory beyond the 20th century. AICAN's first NFT was released on Ethereum digital art market SuperRare in August 2021.

AICAN's most staggering achievement yet? Its 2017 collection was the first machine-generated work to pass the Turing test at Art Basel. When human subjects were shown AI works mixed up with works from abstract expressionist masters and contemporary artworks, they were not able to distinguish between works of the algorithm and those made by human artists. Now to put it in context, almost no machine has ever passed the 65-year-old test where a computer must communicate indistinguishably from a human. Not Sophia the robot or Suri – only chatbot Eugene Goostman.

CAN DATA BECOME A PIGMENT?

Expressing how machines think and feel (even dream and hallucinate) is another popular focus point for media artists like Refik Anadol. Imagine if you can put together every single available photographic memory of New York into one fluid artwork. Refik's team used a GAN algorithm to scour the Internet for publicly available photos of New York City, gathering a whopping 213 million images – the largest dataset ever created for an artwork. With the 'poetics of data', the result was a bewitching 30-minute-long Machine Hallucination movie that predicts (hallucinates) new images, allowing viewers to step inside a dreamlike vision of both an old and a future New York.

In another example, Refik collected all the performances ever delivered at the LA Philharmonic orchestra and the WCDH. Working with the Artists and Machine Intelligence programme at Google Arts and Culture, 77 terrabytes of digital memories dating back one hundred years were parsed and turned into data points. These were then categorised by hundreds of attributes, reshuffled and then projected onto Frank Gehry's Walt Disney Concert Hall. If ever a building could dream, this was how it was brought to life between man and machine. IBM's Watson was also 'inspired' by Antoni Gaudí to create a jaw-dropping installation for Mobile World Congress 2017. Watson was trained to detect patterns and trends as it was fed documents, song lyrics, and historical articles about the legendary architect. Using Watson's insights, the creatives at SOFTlab then created the sculpture's framework, which would also move in real-time with Watson's 'tone analyser' as it extracted the mood from tweets at the event.

Currently, humans are still very much in control of the creative process, training AI to collaborate in their artistic visions. Taking thousands of photos of tulips in bloom, UK artist Anna Ridler famously trained AI to generate videos of thousands of the flowers blooming, controlled by fluctuations in the price of bitcoin. Sougwen Chung has trained AI on her own drawing style, then co-creates with a robotic arm alongside her.

The exciting news about AI art is that it's becoming so accessible. AICAN has built Playform to make AI available to all artists – no coding experience required. Google's DeepDream lets artists create trippy algorithmic paintings through a process called inceptionism. There is also Google's more pedestrian Arts & Culture app that lets you transform photos into the style of Vermeer, Kahlo, or Van Gogh. Not forgetting the exciting potential of Dall-E – the latest release by Elon Musk-backed OpenAI. This neural network is trained to create images from text captions for a wide range of concepts. For now, you can command it to make silly things like a carrot-shaped penguin or an axolotl hugging a turnip, but one can easily imagine a scenario where you can give it much more complicated written instructions. Even a whole animated storyline?

If you scroll down the list of founding members of AIArtists.org – a global clearinghouse and the largest community of AI artists in the world – you'll see that it's largely dominated by visual artists. According to artist and sceptic Mario Klingemann, it's perhaps because our eyes are much more forgiving than our ears. This is exactly the reason why OpenAI decided to tackle music for its Jukebox AI model – 'because it's hard'. Its first attempt does a pretty good job in generating genre-specific music in the style of specific artists. Want to hear Céline Dion sing *No Diggity*? Can be done. Billie Eilish singing reggae? Sure. Want Elvis to sing your own composition? That's also possible. Give it just twelve seconds of *Seven Nation Army* and the result is not wonderful, but quite surprising. Especially if you consider that it matches music and voice and suggests snippets of lyrics. It just goes to show that all you need is a good riff. No look into AI art would be complete without a mention of deepfakes. Vocal Synthesis, a YouTube channel dedicated to audio deepfakes, has brought up an interesting debate in terms of usage rights. It uses AI-generated speech to mimic human voices, synthesised from text by training a state-of-the-art neural network on a huge amount of audio samples. Some of the videos are silly and fun, like Bill Clinton reciting *Baby Got Back*. Others, like Jay-Z rapping Eminem's *Lose Yourself*, was so realistic that his label Roc Nation filed a takedown order, claiming that the content 'unlawfully uses AI to impersonate the client's voice'. The claim was dismissed – maybe because AI impersonation can be seen as innocuous as human impersonation. Or perhaps since Google itself successfully argued in the case of 'Authors Guild v. Google' that machine learning models trained on copyrighted material should be protected under fair use.

Is AI music mainstream yet? Depends on your definition, but absolutely. In its simplest form, Spotify uses AI daily to sift through thousands of newly loaded tracks to suggest the most popular ones (probably not based on the fairest data sets). Tech companies are putting big investments behind AI composing and music creation. Sony has created Flow Machines, an AI system that has released the songs *Daddy's Car* and *Mr Shadow* plus a whole album – *Hello World* by the music collaborative Skygge. Google's Magenta project, an open-source platform, has helped produce songs like the first-ever pop album by Taryn Southern. The YouTuber also used tools from the IBM Watson suite, Amper, Aiva and many more.

Aiva is mostly known for classical music composition but has recently branched out into rock with the song *On the Edge*. By learning and interpreting music from Bach, Mozart, and Beethoven, Aiva creates soundtracks for game studios, film directors and ad agencies. The AI virtual artist has released an album called *Genesis* and is the first AI ever to officially acquire the worldwide status of 'Composer'. It now owns copyright under its own name, registered under the France and Luxembourg authors' right society (SACEM). In another first, tech startup Auxuman released an album created entirely by artificial intelligence. 'Vol.1' is the work of five artificial intelligence 'artists': Yona, Hexe, Mony, Gemini and Zoya.

POWER TO THE MACHINES (AND THE PEOPLE)

As with AI art, there are many options for the amateur musician, from Popgun to Amper (owned by Shutterstock) to Jukedeck (owned by TikTok). The openness of these programmes gives untrained musicians the opportunity to express themselves with sounds and beats our ears can't begin to comprehend. It's not just music producers and DJs who are using machine learning to mix new sounds. Just as Instagram levelled the playing field of photography, so too will AI democratise music production and distribution. Soon, everyone will have the modern-day version of the Madwaves MadPlayer at their disposal. Released over 20 years ago, the MadPlayer was a pioneer in making AI-assisted composing mainstream. With a bank of over 600 instruments and sounds, MadPlayer's unique Generative Music Algorithms (GMA) helps you mix endless tunes in any genre – from techno and ballads to R&B and house. The use of algorithms to create music is nothing new. In the 1950s, computer programmes were used to generate piano sheet music and in 1958 Iannis Xenakis famously wrote the first algorithmic piece ever (using Markov chains). Generative music is well-trodden ground. Brian Eno's Bloom app is over 10 years old – the original '21st-century music box' that can create an infinite selection of compositions and visualisations to match. French composer Jean-Michel Jarre's Eon app also produces a constantly evolving suite of music. Where it's different is that each time you play the album, it generates something special for you.

It varies for every individual – truly personalised music just like Björk's AI-powered composition, Kórsafn. This lobby score created for the Sister Hotel in New York uses Microsoft AI and a rooftop camera to record and translate sky activity (planes, clouds, birds) into data. From this data, the AI programme creates a unique symphony with snippets of Björk's choral archives and recordings by the Hamrahlid Choir of Iceland. As the seasons change, as the sun rises and sets, as birds flock to the city in springtime, the AI learns about new and different weather events, and adapts these influences in the music.

MUSIC THAT HEALS

Music that follows the nuances of the weather or the movement of birds – what's more beautiful than that? Personalised audio has heaps of exciting potential – something that Samsung will bring to market soon with its Generative Soundscape. This is no ordinary sound system – it will analyse your home environment and use AI to create real-time ambient sound for work, rest or sleep. Music for wellbeing is going to be huge over the next few years. A Samsung UK survey discovered that, compared to 2019, roughly a quarter of millennials are now listening to over five hours of music a day and more than 50% of participants cite music as their number one 'feel-good' source. First there were meditation apps; soon big-name stars will sing adult lullabies or create anti-anxiety albums. Grimes (aka Claire Boucher) has already collaborated with Endel – one of the pioneers in generative sound technologies.

By capturing biometric and situational data, Endel creates an ever-changing sound environment to energise or relax your brain and body. Endel doesn't call their soundscapes music; they see it more as an ambient 'sound blanket, almost like adding another biological function as unconscious as breathing'. When you think of music as something therapeutic, it's not that hard to imagine soundscapes for calming you down, lifting your mood, or curbing your appetite.

Even the way we listen to music is changing, with apps like Audible Reality promising stunning immersive audio through AI-driven, 3D audio enhancement. We haven't touched on vocals yet. With Yamaha's Voicaloid, you can create songs just by inputting lyrics and a melody. From its many Voice Banks, you can cherry-pick a voice to match your style – recordings of actual human voices, for example 'Amy: Female English' or 'Ken: Japanese Male'. By using deep learning to analyse singing traits such as tone and expression within singing recordings, Vocaloid can also synthesise singing with any melodies and lyrics, matching unique mannerisms and nuances.

Hatsune Miku is one of these singing voice synthesisers featured in over 100,000 songs. Her name means 'the first sound from the future' and if you consider her meteoritic fame, it certainly is a taste of things to come. She has over two million followers on Facebook, has opened for Lady Gaga and performed from LA to Europe. Since she was released in 2007, her fandom has exploded. Her image is licensed out for all sorts of merchandise, games and products, and she is by far the biggest J-pop star of all time.

Back in 'real' life, Yamaha also used Vocaloid to reproduce the singing of the late singer Hibari Misora for a brand-new song, released in honour of the 30th anniversary of her passing. Actual recordings of the artist's songs and speech made while she was still alive were used as machine learning data to reproduce her singing.

PROBABILITY

SCIENCE

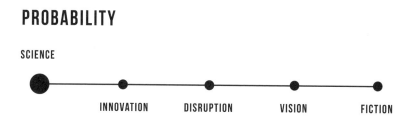

INNOVATION   DISRUPTION   VISION   FICTION

You might as well get the popcorn – AI art and music are here to stay. On the music front, it's a buddy to jam with, a powerful composing tool, and an endless source of inspiration. It doesn't have any taste, but it learns fast. With GPT-3, we already have the lyrics sorted. They may be a bit formulaic now, but imagine the creativity by the time GPT-20 comes out. With Vocaloid, synthesised singing is here, and very soon, probably virtual choirs and bands singing beautifully together. And of course, the actual instrumentals and music mixing will be enhanced further and further until they can create symphonies and scores on their own.

With art, perhaps things are a bit more subjective. Is AI art really art? Who is the artist here? Who deserves the ownership? The definition of authorship will start many debates. If the artist is the one that creates the image, then that would be the machine. If the artist is the one with the artistic vision, it's the humans. Right now, humans are driving the artistic vision but that may well change fast. For Refik Anadol, data is his material and machine intelligence is his artistic collaborator. Together, man and machine will do magnificent things.

Ahmed Elgammal, the developer of AICAN, compares AI art to photography – an art form that was initially dismissed by tastemakers. When photography was first invented in the early 19th century, it wasn't considered art. After all, a machine was doing much of the work. Critics eventually relented. A century later, photography became an established fine art genre. Perhaps art produced by artificial intelligence will go down the same path.

The argument used over and over about whether robots will take our jobs is that they are simply not creative. Still worried that painting will be 'dead', as the French painter Paul Delaroche pessimistically declared in 1840? We'll let artist Mario Klingemann have the last word. 'In the end, competition always forces us to get better. To see what makes us as humans still special.'

THE MOTHS
ARE IN THE PESTO

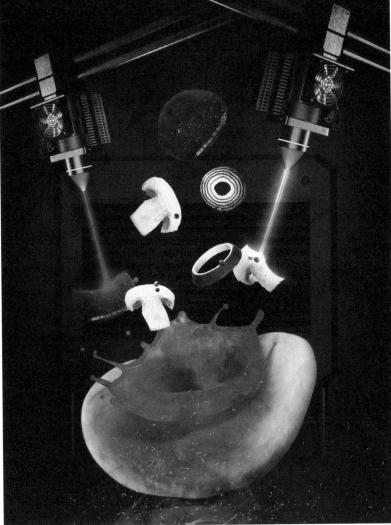

Friday night in the Brown household means two things: Pizza Fridays and watching the latest episode of Neurochef together. After a week of only sipping on specialised meal replacement drinks, everyone's excited to chew on something different. Mom Lisa and teenage daughter Cassie are prepping the food in the kitchen with their fancy new Pizza Popper machine, which 3D prints and cooks perfect garlic flavoured pizza bases in two minutes, then proceeds to print each family member's order for toppings. Ten-year-old Bruno loves spicy green spelt meatballs, fried yam and mozzarella, while dad Fariq sticks with his old double-meat favourite of fungi pepperoni, mealworm mince and pickled cactus. Cassie is trialing a new jellyfish-only intermittent fasting diet, so she's prepping her pad thai style noodles on the side.

There's huge excitement in the air for the semi-final of the world's favourite cooking show. Three teams will have a final cook-off for a place in the final, judged by top neurogastronomists. On the team: a talented chef, scent DJ, body music composer, plate curator and senseuse (sensory stylist). Neurochef is based on the pillars of neurogastronomy – the science of how our brains 'taste' food through our senses of sight, smell, touch, taste and sound.

Contestants are required to prepare a showstopper three-course meal with unique challenges. No seasoning is allowed so they'll have to conceptualise special sounds and aromas to trick judges into smelling and tasting mouth-watering creations. If one team draws the red card, they get to sabotage an opponent's creation by restyling their plates. The Brown family is on the edge of their seats to see if their favourite team's food gets the dreaded red plate (known for switching someone off their food). The senseuse will have only sixty minutes to come up with a virtual reality experience to immerse judges (and everyone at home) in a cozy or exotic environment that compliments the chosen cuisine, while telling a story. It's food theatre deluxe.

The winners will each walk away with an electric self-driving car and a year's contract with personalised nutrition giant, Tummible, to work with AI recipe writers on fresh new food plans. Pizzas in hand, the Browns are glued to the screen to see what happens next...

PERSONALISED NUTRITION WITH AN EXOTIC TWIST

By 2051, food will be far more of a fuel than a fixation (although, for some, still very much a TV obsession). By then, we will be so sophisticated in how we bio-hack our bodies that we may not eat much at all. With connected health devices tracking our every step, breath and heartbeat, we'll be relying on delicious custom-mixed meal replacement drinks and powders to turn our bodies into high-performance machines. Those who live in first-world countries will be government-issued with free fitness trackers to incentivise healthier living. Those who can afford it will subscribe to precision food prescriptions, fine-tuned through microbiome mapping.

Engineered with a dizzying array of proteins, vitamins and minerals, essential omega fatty acids and fibre, you'll get exactly what your body needs every time – nothing more, nothing less. A combination of glucose implants and metabolic breath sensors will tell you exactly how your body is functioning in real-time. Conditions like obesity, diabetes and high cholesterol will become a thing of the past.

If you do decide to eat food, personalised nutrition will guide your every choice. You'll know exactly which foods will cause cravings and how each bite influences your metabolism. Everyone will be on the metabo-friendly band-wagon, from personal appliances to grocery stores, restaurants and fast-food outlets. Your app will sync with your smart fridge to recommend and order food based on your personalised diet plan. On supermarket shelves, where you once only had options like low vs. full fat or salt vs. no salt, foods will be available in lots of variants customised to common metabolism types (e.g. high fat, high protein, medium fibre).

Speaking of grocery stores – where have the meat and fish aisles gone? They are still there, just not how you remember them. Gone are the days of your fishmonger filleting your trout in-store or a butcher cutting your rib-eye steak to size. You can still pick your perfect meat cut but it will be shaped like a lamb chop and lab-grown from animal stem-cells. Or your fish fillets will be grown from fungi combined with algae and plant-based ingredients. Each fresh product will also come in metabolic-friendly varieties that include just the right ratio of fat, protein, minerals and vitamins you'll find in the real deal.

You could probably still get hold of meat or fish from niche suppliers but it will be somewhat frowned upon – like smoking or eating foie gras – not quite illegal but certainly not mainstream.

Next to the 'fish' and 'meat' counters, you'll pore over the week's selection of edible insects to include in your Bolognese sauce or rich chocolate cake. These will be sold unprocessed (wings and legs on) for snacking and cooking, or available in a large variety of familiar shapes, from fruit loop cereal and couscous to burgers and sausages. If creepy crawlies don't tickle your taste-buds, you'll feel like a hero for shopping in the 'harm-to-table' section: where invasive species like crayfish, crab and jellyfish help foodies-in-the-know dish up a greener meal.

Remember all those aisles filled with carb-rich things like pasta, rice and flour? These will now be largely protein-based offerings, with everything from cookies to crackers enriched with protein from sources like desert-grown seaweed and fruit flies. You'll probably have a hard time finding the basmati rice or durum wheat pasta amongst the packs of cowpeas, finger millet and amaranth, and bags of Bambara, Marama and Adzuki beans.

SMARTER EATING MEANS HEALTHIER BODIES AND EARTH

Chargrilled burgers with melting cheese, creamy fish chowder, ravioli with melt-in-the-mouth fillings. In 2051, all your favourites will still exist – albeit in completely different formats. As you tuck into your lab-grown chicken burger on a protein-rich fonio flour bun, you'll listen to a special neurogastronomical soundtrack to make you feel satiated faster (no deep bass hip-hop for you). Your ravioli will be filled with smoked carrot salmon, chopped caterpillar and sharp fungi cheese, while your tofish chowder will be thickened with insect milk, and swimming with delicious invasive crab legs. You'll sip this with a textured spoon – proven by neurogastronomists to make foods taste saltier without any added sodium. Portions will be tiny as you'll have conditioned yourself over the years to eat less and less. If you're still hungry, you'll finish your meal with a dandelion root and tiger nut milk latte, then immediately start tracking in real-time how this food is metabolised.

Our palates will become more adventurous as food scarcity compels us to try different things. We may not know what teff, ube, rabi or enoki are right now but in 2051, these future foods will be very much part of our day-to-day cuisines. Why? Well, firstly, we've been eating far too boringly for far too long. Did you know that 75% of the global food supply comes from only 12 plant and 5 animal species? And that 3 types of plants (rice, maize, wheat) make up nearly 60% of calories in the entire human diet? This is not just narrow-minded eating, it's also missing out on many valuable sources of vitamins and minerals. However, we won't be eating strange things for novelty's sake. Our collective conscience will catch up with us because we can't continue destroying the earth for what is essentially a luxury diet. How can we keep munching away at chicken and beef if 80% of the world's agricultural land is used for animal farming, which produces just 20% of the world's calories? What's more, the world simply can't sustain food production for the 10 billion people expected to be living on earth by 2050 unless farming and food industries become much more sustainable.

One way we're already doing this is by going meat-free, and continuing to investing in new meat and fish alternatives. The hottest topic of all? Entomophagy – the custom of eating insects. Snacking on bees, bugs, moths and worms is nothing strange in places like Thailand, China, Brazil, Mexico and parts of Africa. With more than 2000 species of edible insects in the world, there is much to explore. Crickets, grasshoppers and mealworms are particularly popular for being high in protein. Salted butter mealworm snacks, anyone? Fig, chocolate and buffalo worm powder protein bars? French manufacturer Jimini has been on a mission to normalise insect eating since 2012. Eat Grub, the forward-thinking UK insect-eating evangelist, is one of many to have brought out a cookery book and offers a range of freeze-dried, ready-to-cook insects, cricket powder energy bars and roasted grub snacks.

EntoMilk, a dairy alternative made in South Africa by blending the larvae of the black soldier fly, is used for luxury ice cream. Micronutris was one of the first French companies to farm insects. Today, they sell a wild array of insects to use in cooking, with delicious recipe ideas to match like pumpkin soup with hazelnuts and crickets, or ravioli with mealworms and chestnuts. In Switzerland, Essento's insect burgers and meatballs have been on sale in Coop, one of the country's largest supermarket chains, since 2017. Another non-meat option on the rise is lab-grown meat, with companies like Meatable and Ivyfarm leading the way in making slaughter-free meat. Just like taking a cutting from a plant, Meatable takes a sample of cells from an organically raised animal and nurtures it in a large tank called a bioreactor. In under three weeks, these cells multiply to around 1kg of meat. Ivyfarm will take its first product (pork sausages) to market by as early as 2023, with Angus beef burgers and Wagyu meatballs to follow. This production method has the potential to use 96% less water and 99% less land than industrial farming. Somewhat less extreme will be the next generation of vegan 'Beyond Meat' plant-based meat. Like your meat on the rare side? Nothing is off-limits for these entrepreneurs – Impossible Meat even has a burger patty that 'bleeds' just like meat. We'll continue to experiment with meat-like plants, such as jackfruit for pulled pork and banana flower for fish, to slowly wean ourselves off meat-eating.

Now imagine you can take all these interesting new superfoods, grind them down and ingest them all at once? Imagine you can find the perfect fuel for your unique body – for example, the exact potion for your IBS, your weight loss and your gluten intolerance. The next generation of meal replacements is already doing just that. Mana's 'subatomically perfect' complete meal solutions contain 42 essential nutrients, with protein and fibre from soy, peas, oats, hemp, rice, and algae. The list of these 'Silicon Valley' drinks goes on, from Soylent and Huel to Yfood, OWYN and Saturo. Every one of them shares the same idea – that food can be simplified for the better. For too long, we have optimised food for taste instead of for its primary purpose of providing all the nutrition your body needs. These drinks and bars bring precision and efficiency to how we feed ourselves. Dutch brand Queal goes a step further, with a dashboard to track the impact Queal use has on your health, wealth, and environmental footprint. Adding a gamification element, you can also compare results with friends, earn points and level up.

It's a given that personal nutrition, wearables and connected devices will rule culinary choices in thirty years' time. Apps and algorithms will identify what we should eat or avoid, and will know exactly what we keep in our smart fridges or order in our online shopping carts. Metabolic health doesn't just influence your weight; it's also responsible for how well your body processes food (i.e. gut health), your energy levels and how well you sleep. No two people are the same when it comes to metabolism. Two people of the same age and health vitals can eat the exact same diet, but have vastly different body shapes and health conditions. This is something that Israel's Weizmann Institute of Science has been studying for years. One of its initiatives, Project 10K, is collecting personal health data from thousands of Israeli residents, which will be processed using sophisticated artificial intelligence tools to identify a person's likelihood of disease. They are already using algorithms to diagnose Covid from lung images, predict the risk of gestational diabetes or determine someone's risk of developing an acute form of leukemia.

Some of the startups already offering bespoke nutritional advice by algorithm include DayTwo, Thryve, Nutrino and Zoe. Simply send in some blood and stool samples, wear a continuous glucose monitor device, and do some home tests and questionnaires. A few weeks later, you'll receive a detailed report comparing your results to those of thousands of other people. If you want to go a step further, invest in a breath analyser like Lumen – the world's first portable device to accurately measure metabolism. In just one breath, Lumen will measure the CO_2 in your breath, which indicates the type of fuel your body is using to produce energy. With Lumen's daily insights, you will improve your metabolism by improving your body's ability to shift efficiently between using fats or carbs as a source of energy.

PROBABILITY

SCIENCE

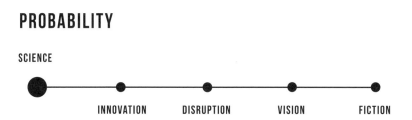

INNOVATION   DISRUPTION   VISION   FICTION

Thirty years from now, the word flexitarian will take on a whole new meaning. Will we flex eat between meal replacements and real food eating? Or between plant-based and lab-grown? Will you be vegan with a side of insects? Will your pescatarian diet mostly consist of jellyfish, farmed fish and algae products? There will definitely be terms for our new metabolism- and earth-friendly way of eating.

Our bodies are fascinating things – rich with data waiting to be mined by everyone, from eager start-ups to tech giants, like Amazon and Apple. Our tongues might 'tell' our brains we're tasting one of five things, but taste is very much influenced by how we see, hear, smell and feel our food. And then there is how flavour is perceived by our gustatory cortex (the perceptual taste centre of your brain). Studies from Columbia University have already proved it possible to activate the bitter-responsive neurons of a mouse's brain, so that it disliked the sweet water it was given. In which case, it could be entirely possible to trick human brains to like lettuce over donuts in thirty years' time, right?

Hippocrates predicted over 2000 years ago that food should be our medicine. While this still rings true, let's hope that it will continue to be our way to celebrate togetherness and nourish ourselves, not the intense and somewhat unhinged obsession we have with it right now.

THE CASHIER IS DEAD

Dieudonné sips his morning coffee – an Ethiopian blend made with caffeine-free beans recommended to him by the RoboDietician in aisle 4 – and looks at the giant screen in front of him. As the general manager of a 4-storey gigamarket in the centre of town, he loves to start his day with lots of green on the dashboard, as it means his work will be a breeze – at least until lunchtime. That's not the case today. There is a bit of orange blinking in the far-right corner of his screen and an ugly red dot smack dab in the middle. He taps the keyboard of his HoverChair to bring it closer to his desk, where the brakes automatically lock the perfect distance away from his screen.

Dieudonné fights the urge to roll his eyes. He knows from experience that the orange dot signifies a problem with the salmon supply – again. When will shoppers finally understand it's 2051, salmon is an endangered species and the quotas for fishing it are lower than ever? With a few quick clicks on the dashboard, he activates the automated solution from his Price Management System. An AI-generated promotion on haddock is instantly displayed on the in-store screens, incentivising shoppers interested in salmon to buy another type of fish for supper. An automatic notification is then sent to the RoboDietician in the seafood court to let her know of the shortage; she will adapt her proposals to shoppers accordingly. And of course, the electric price tags for both types of fish in the fridges are immediately changed as well. Later this evening, before going home after closing time, Dieudonné will – out of habit – check if these products' prices have been altered on the outdoor LCD screens and vending machine for late night shoppers.

With the fish issue quickly solved, Dieudonné now turns his attention to the red dot. He reads the description of the problem. There is no automated solution possible. This is why he still comes to work every day, for things only a human can solve. Using the keypad once again, he programmes the HoverChair to go down to the store. Then he spins around and heads for the lift.

Gliding out on the ground floor next to the gigamarket entrance, Dieudonné looks to the left. The store's only two tills are manned by humanoid robots that pack groceries and pretend to smile for clients who need the interaction.

Dieudonné misses his favourite cashier, Miss Imani. She was so warm and friendly, always helpful. When she retired, planning her goodbye party was hard. *Replacing her with a humanoid made more sense financially, as it could do what she did, but also gather and analyse real-time data about shoppers checking out.* This data is now instantly fed directly into the cloud where it is combined with other data from his store to fuel the algorithms in his prediction software.

Dieudonné glides past the fruit and vegetable aisle. His HoverChair swerves rather abruptly for a kid in AR glasses playing a game with an invisible Hello Kitty. The kid's mother is distracted by a conversation with the aisle's Robo-Dietician over the amount of carbs in a banana versus an avocado. Dieudonné doesn't mind. He knows these branded AR experiences always create an uplift in overall footfall of over 23% for the day. He glides on to the dairy fridges. The screens above the fridges light up when he passes, showing how the packaging of perishables has improved in his gigamarket.

A bit further, he passes a senior comparing a piece of lab-grown meat with a vegan alternative by simply holding up the products in front of his phone's gigamarket app. The senior nods, puts the vegan product back and walks towards the wine section. There, a young couple uses the touchscreen table to pair the perfect bottle with the recipe for the meal they plan to cook this evening for friends. One of them is allergic to sulfates, so they fuss over the right wine for the right price. Dieudonné knows that – with the ingredients for the meal already in their cart – the system will recognise this and reward them with a personalised coupon for an entire carton of sulfate-free white wine nearing its 'best-by' date.

Finally, Dieudonné arrives at the cause of the red dot on his dashboard: the frozen goods section. The dashboard had warned of a spike in energy usage in this area. He checks if all the doors of the freezer section close properly, then looks for puddles that might indicate a broken part somewhere. It takes him until lunchtime to figure out and manually fix the problem. When he returns to his dashboard, all he sees is a satisfying sea of green on the screen.

SUPERSIZED MEGAMARKETS WITH AUTONOMOUS AI AND ENTERTAINMENT IN AR - REALLY?

For the immediate future, technology will not hugely impact our everyday experience as shoppers. But make no mistake, it will be there, embedded into the workings of our favourite grocery stores. Due to advances made in AI and Machine Learning, footfall forecasting, dynamic pricing and supply chain management will be highly automated, requiring only a single person to oversee the process on a central dashboard.

Soon, self-checkout will be the norm. Most tills will disappear, with customers automatically charged as they exit the store. Some retailers will still incorporate robot-cashiers for a more human experience, and for customers who prefer the feedback of a once in-flesh-cashier repeating the total cost of the groceries and wishing them a good day before they exit.

This will make cashier jobs a thing of the past but will also make space for new vacancies. Pricing of goods will be determined and updated instantly and adapted automatically across multiple channels. The most popular products will be continuously on display, made possible only because real-time AI collects information on supply and automatically regulates orders as stocks run out. Data-based predictions will become imperative, like determining seasonal product demands based on last year's data – in combination with data from social listening (scraping data around ingredients and recipes, for example). With all these tasks automated and autonomously performed, human store managers will evolve into market researchers and adverse event agents. They'll be paid to react fast whenever there's an unpredictable occurrence, like a sudden major flooding of the premises due to rain.

With efficiency of supermarkets at an all-time high, attention will turn back towards increasing the number of products per basket. In each food aisle, we may encounter a dedicated food expert, or dietician, who will help us towards product choices best suited to our needs. Less privacy-cautious shoppers and those pressed for time will simply press their smart watches against a touchscreen table for a better insight into their individual health and conditions – with personalised shopping proposals and promotions immediately offered. With a simple click, a shopping list is sent to their phones and the corresponding coupons sent to their digital wallets.

Shoppers will not only be encouraged to spend more money but also more time in their favourite supermarket. The key to this is the enhancement of the shopping experience. One concept that is increasingly gaining traction is to bring amusement-park activities to the supermarket space. Upon entrance, much like they were once offered barcode-scanning devices, shoppers will be offered AR glasses that will fill the mundane supermarket environment with colourful characters and animations. These AR glasses will also bring the in-store data collection to a brand-new level, as they'll offer the constant influx of data on retail analytics through 3D eye tracking.

Instant gratification and exceptional experiences will play a major role in the popularity of supermarkets. At first, they might feel gimmicky, but after a while, they will really focus on personalisation and relationship building. Then, supermarkets will realise they need to play a role outside of their proper ecosystem – collaborate with mobility providers, healthcare providers, entertainment providers and brands.

Take the example of a brand like *Hello Kitty*, originally a shoe brand with a cute kitten design from the 1970s. By 2020, it had already branched out very successfully into the respective markets of sneakers, sweets, furniture, even motor oil and car engines – taking it far beyond its original, pre-teen audience. So it's not a stretch of the imagination for a high-end retail brand to put on a 'Hello Kitty Day' in the future: parents go shopping while their children walk around with AR glasses that show the popular feline following them around. This will not only increase the average amount of items in the basket but would also sell out *Hello Kitty* merchandise in no time. Perhaps an added benefit of this could be a function that helps parents to track their children, thus largely eliminating those heart-stopping moments every parent has felt first-hand when their offspring can't be found.

We've all experienced the frustration of needing a specific grocery item – only to find that the supermarket is closed. In 2051, users will obtain whatever they need *whenever* they need it. Store hours will stay the same, meaning that European supermarkets will still be open from 9 a.m. to 6 p.m., and they'll still enable shopping outside of those hours with weekend (and night) delivery service. What will be different however, is the arrival of LCD displays on storefronts. The displays will showcase a range of products with QR codes below them. Upon collecting QR codes of all the products wanted, a customer will pay either digitally or through facial recognition. They'll then select the time at which they want what they bought to be delivered and specify the delivery address.

Storefronts will also have a vending machine function where most sought-after products can be bought instantly. After-hours shoppers will open the app showcasing the real-time contents of the vending machine they want to purchase from. After purchasing the product, they'll get a QR code generated within the app which can then be placed against the vending machine to release their products. It will take just one more pandemic like Covid-19 for us to realise the value – and the need – for storefront shopping, even during working hours. No longer would shoppers need to spend long periods of time in a closed space where the spread of diseases is a given.

It's easy to see how this system would also help large families with big grocery needs. The drudgery of dragging shopping to an electric car or bike to transport everything, could become a thing of the past. With a new, hybrid shopping model, users will be able to go to the supermarket, scan the products with the barcode scanner, and have them delivered to their home while enjoying the entertainment the supermarket has to offer. Shopping carts will still be available for whatever products users need immediately. The rest of the visit will be just for the sake of touching or smelling products, benefiting from personalised health advice and product promotions, and seeing a lifesize *Hello Kitty* juggle some apples and oranges in AR.

The delivery system becomes even more relevant when one considers how our transport habits are set to change over the coming years. According to a trio of cycling associations – Cycling Industries Europe (CIE), CONEBI (the Confederation of the European Bicycle Industry), and the European Cyclists' Federation (ECF) – Europeans are expected to buy an extra 10 million bikes per year by 2030, a whopping 47% more than in 2019. The new 30 million per year total will take bike sales to more than twice the number of passenger cars currently registered per year in the EU. With so many more bike owners expected, supermarkets will have to compensate for the lack of grocery storage space these two-wheeled vehicles have, compared to the spacious boot of an average car.

IT ALL COMES DOWN TO GUTS... AND USER DATA

In 2021 the idea of till-less supermarkets is already beyond its conceptual phase. Actual stores like that already exist. One such example is the Amazon Fresh London Store. Users must identify themselves on arrival by scanning a barcode displayed within their account on the standard Amazon Shopping app. Products are scanned when taken off the shelf and placed in the bag, charged to the individual user's Amazon Account after they walk out of the store. For this to work, powerful technology was already developed. But it isn't limited to creating frictionless user experiences: AI is currently in use for high-impact supply chain management areas, like scheduling, spend analytics, logistics network optimisation and partial forecasting. According to McKinsey, 61% of manufacturing executives report decreased costs, and 53% report increased revenues as a direct result of introducing AI in supply chain.

Optimising the way a supermarket functions is one thing, enhancing a retailer's relationship with its clients is another thing altogether. According to the Food Marketing Institute's 2018 U.S. Grocery Shopper Trends report, 55% of users saw their grocery store as an ally in their wellness efforts – on par with health clubs. A logical next step is in-store dieticians that provide various services – from customised meals and diet plans to food demonstrations. US store Hy-Vee is one of the pioneers in the use of in-store dieticians in supermarkets. In 2018, they even rolled out a dietician-led store tour programme. The effort focused on three conditions: diabetes, hypertension, and high cholesterol.

This relationship between shopper and supermarket can go beyond wellness, and into the entertainment industry. Merging entertainment and mundane activities like shopping for groceries has been imagined by many visionaries and artists already. A perfect example is the *Hyper-reality* movie by Keiichi Matsuda. This movie showcases a world enhanced by a commercial AR layer. It's far from just imagination at this point: Display Plessey Semiconductors Ltd has been developing programmes for AR glasses, making that additional layer feel more attainable than ever. That's not all, 3D eye tracking has already been provided as a service by specialised companies like GazeSense, to conduct user shopping research. It's possible that we'll be seeing this feature incorporated into upcoming AR glasses from Apple or Facebook, with versions of them intended for supermarkets.

All of this shows what a supermarket will look like during the day. At night, stores are quite likely to remain closed due to lower footfall. So, what about more acute needs for specific products? Though smart storefronts might seem distant, the presence of vending machines like Acure Pass indicates that these may be a step closer to becoming a reality. These machines already display products on an LCD screen and make it easy to purchase via an app and collect with a personally generated QR code.

PROBABILITY

SCIENCE

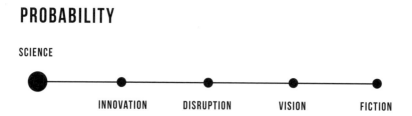

INNOVATION   DISRUPTION   VISION   FICTION

When one considers the technological advances already in existence in the retail space, it's easy to see what lies ahead. Our vision is not reliant on the development of new technologies, just the courage to combine existing ones in new and exciting ways, and in doing so, reimagine the way we think, and shop.

Supermarket retailers have nothing to fear. We will always need food, drinks and clothing. And because of this, there has perhaps been some complacency around shopper experience in this category historically. But this need not continue. AI is here to stay. Innovation is inevitable. It just remains to be seen which retailer giant will dare to pioneer the creation of the ultimate, AI-driven gigamarket of the 2050s, filled with compelling AR entertainment for its users. Let's go shopping!

THE UNCANNY VALLEY

STARRING

HEATH HOFFMAN   SOUTH WEST

"Nomination Best
META ACTRESS
leading role 2051"

Lake Kolyma

COMING SOON

Day 25 of shooting Lake Kolyma, in the year 2051. The director of photography, Joel, is watching a scene on his VR glasses. Actress South West glides across an eerie landscape. She is playing the part of a two-headed tribal leader who has fallen in love with an intrepid explorer, played by actor Heath Hoffman. A 1000-strong crowd is cheering them on as they fight-dance in an epic showdown of their love. It's safe to say that romcoms will be somewhat weird in thirty years' time. The setting – an abandoned mining town in Kolyma, Russia – has lush forests, snow-capped mountains and a frozen radioactive lake. Buildings, transformed into post-apocalyptic Snow-White fortresses, are overgrown with animated thorns.

The director, Margot, yells 'cut' ! The acting is locked but she wants to make a change to the background. She motions to Joel to move a mountain on the horizon. He huddles with the technical team to make it happen, which takes about an hour. The delay could irritate a diva actress but it's no bother to megastar South. Why? Because she's not even there. The actress you see on the screen is a synthesised version of North, created in a Hollywood backlot a few months ago. Her co-star? A meta-human mash-up of the iconic deceased actors Heath Ledger and Philip Seymour Hoffman – as voted for by the audience of the streaming channel. Fans also voted on the costume the actor is wearing, which doesn't quite fit with Margot's vision but is good for marketing.

We may be moving mountains in 2051 but that's not what's happening here. When Joel takes off his VR glasses, all he sees is an enormous circular LED wall. On it, we see a super-realistic projection, complete with crowds of people (a mix of extras and meta-humans). Only some placeholder stand-in actors and key props are there for scale in the studio – the rest of the magic happens on a massive interactive holodeck.

When they wrap for the day, Margot sits down with the scripts for the second series – created by an Oscar-winning script bot, Mollywood. Known for her intense family dramas, this will be her first foray into adventure comedy, so there is a lot of hype around it. Soon, she finds herself laughing out loud. The humour is razor-sharp and remarkably on point for an AI-generated script.

Meanwhile, Joel heads into an editing suite to oversee the live rendering of the next locations. The plot sees characters time travelling to a tropical destination, so a team has been dispatched to Tristan da Cunha – a group of volcanic islands in the South Atlantic Ocean. It had an eruption earlier that year, plus it's so incredibly remote that it wouldn't be cost-effective to take the crew out there. Luckily, it's no longer necessary with specialised equipment that captures scenes in 3D down in pixel-perfect detail. The small team is touring the remote island, capturing specific mountain and beach scenes while the shoot is happening. Margot wants a bigger mountain, so a bigger mountain she'll have by morning – captured and rendered in just 24 hours. If that doesn't work, there is always a vast library of lidar (light detection and ranging) scanned locations from every corner of the world to choose from. Lidar scanning of Tristan isn't possible because of the mirror town recently built by an NFT billionaire. 3D volumetric scanning has advanced a lot but still can't deal with objects like mirrored surfaces.

WELCOME TO A WORLD WHERE (LITERALLY) NOTHING IS REAL

Filmmaking in 2051 will be in the hands of a small group of creatives, and powerful virtual production software with unimaginable processing power. Where before millions of dollars were poured into moving hundreds of crew to exotic locations abroad and huge set builds, everything is now in-camera at once.

Locations and props will be hand-picked from content libraries, offering an endless supply of textures, animals, furniture, foliage, vehicles and more to blend into the scene in real time. No more working in foreign locations at the mercy of weather or daylight shooting hours. No more spending months in post-production. No more expensive re-shoots. No more movie flops because the director couldn't see it all coming together until it's sadly too late. Assets are created once and used repeatedly. Films and TV shows can now be created on more flexible schedules, with actors working on several projects at once. Talented creatives will design and 3D print imaginative props, while smart costumes will be programmed to display specific textures to boggle the human eye.

Casting will also take an interesting turn. Actors will now film a few key scenes, and their digital alter egos will step in to finish the job. If budgets are tight, they don't need to be present at all. Deepfakes – machine learning systems that can learn to exactly mimic the data you feed them – will have progressed so far that it will be quite possible to create super-realistic facial overlays. Actors can be recreated posthumously, or filmmakers could decide to use deepfakes to create new actors out of 10 different faces. Audiences will be involved with content creation – just as the NewNew app polls followers to tell influencers what to say, wear or do. Writers will still create initial scripts but bots will finish them off.

It will be a world where almost nothing is real unless it has true artistic meaning. A world of wonderful efficiency and collaboration.

VIRTUAL PRODUCTION IS EVOLVING FAST

One of the biggest stories of 2020 in film and gaming circles was the revolutionary technology used to make the Star Wars spin-off, *The Mandalorian*. A collaboration between Industrial Light & Magic and a real-time game engine by Unreal, it was the biggest invention since the green screen. This monster LED screen (now commercially available as Stagecraft) is the largest, most sophisticated virtual filmmaking environment ever created.

The virtual production unit measures 20 feet tall and 75 feet across, offering a curved 270-degree cinematic landscape. Before this, either giant sets were built, or actors behind CG characters would play out their roles in front of a green screen. Now, lighting, sound, location, VFX and acting all come together in the moment. Soon, the entire workflow will go real-time, with less time waiting for computers to process. Even sooner, there will be LED screen set-ups from Cape Town to Mumbai. Unreal is already marketing its gaming engine beyond filmmaking – architecture, broadcast and live events, training, automotive and cultural tours are some of the industries it's focussing on.

As creatives were getting excited about the opportunities this virtual environment presents, Unreal unveiled another jaw-dropping innovation – an app that can build lifelike virtual humans for games and movies. Called MetaHuman Creator, it makes face customisation as simple as a Sims video game. Skin complexion, wrinkles, stubble and freckles – it's all there, down to each creepy detail. These digital humans are ready for their close-ups. They are very close to conquering the so-called 'Uncanny Valley' – where human replicas that look almost like real human beings elicit feelings of eeriness and revulsion. Film purists like Quentin Tarantino will argue that synthesised actors will never give you the raw emotion or the Jim Carrey-esque facial acrobatics of a real-life actor. It will be the 'death of cinema', they'll protest – just as Tarantino did in the digital versus film debate ten years ago. But at $70 a minute, film is expensive. It probably won't be long until he too succumbs to digital filmmaking. Perhaps even digital actors.

Speaking of production costs, another huge expense is of course shooting on location. This is where lidar technology is bringing exciting advancements, fast. Lidar is becoming increasingly popular for creating realistic computer-generated imagery and visual special effects. Traditionally used for mapping in industries like civil engineering, mining and transportation, lidar is now used to scan buildings or even entire cities in 3D for visual effects. The HBO series *Game of Thrones* relied heavily on lidar to create and recreate their sets. The old city of Dubrovnik was rendered in 3D as the model for the fictional city of King's Landing.

Thirty years from now, we may well have most of the earth's cities mapped out in this way. Perhaps drones could capture scenery inside a volcano, a rain forest, underground cave, or mountain range?

PROBABILITY

SCIENCE

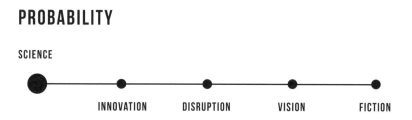

INNOVATION   DISRUPTION   VISION   FICTION

The one thing that won't change: talent will always have a place in this world. When researching this article, I came across a fun 2004 interview from the future where Julia Roberts 'reflects on acting with her synthetic younger self' in the year 2022. While some of the technology sounds plausible (a suit with thousands of location sensors, lifting a 'skin' from her *Pretty Woman* performance), we also know that the legendary actress has nothing to fear from digital actors. And there is no reason to use digital stand-ins with the constant influx of incredible acting talent.

People love people. Real people to whom they can relate. People that age like they do and portray the emotions they feel. They love the versatility of an actor playing a lawyer, then an addict, then a dragon slayer. They love their messy private lives and dodgy accents. Digital actors are boring and will never take centre stage.

But, perhaps actors should be less worried about losing royalties, and rather capitalise on their performances. If digital art can be sold as NFTs, why couldn't an actor sell iconic facial expressions that can be mashed up with meta-humans? Perhaps everyone should be less wary of digital versions 'taking over' and more excited about the endless creative opportunities that lie beyond the uncanny valley.

ATTENTION IS
THE NEW BITCOIN

The day before her retirement, the last day of the school year of 2051, Katinka fires up her school laptop for the last time. She neatly arranges the concentration headbands by the entrance, so early arrivals can grab one in case they forgot their lenses at home. A convenient excuse she has heard so many times, and in so many different shapes. She remembers the start of her career, back in the days when excuses were used for not handing in written homework or 'forgetting' to attach their homework to an email. Now, towards the end of her career, it's the performance measuring technology that is always 'forgotten' by at least one of her students.

It comforts her to know that no matter her age, kids have always been kids. For a while, especially at the dawn of perfotech in schools, she was afraid children would become robots. With their attention monitored non-stop, Katinka feared they would feel under continuous pressure to perform. But she had misjudged the situation (mostly because she herself didn't get her first mobile phone until she was 18 and didn't fully comprehend children's ability to adapt to new tech). The children she had been teaching lately have grown up with technology in an inclusive way, as if trackers and algorithms were a natural extension of themselves. Monitoring their brain performance in graphs and numbers was not that big a leap from what they were already used to in their VR game performance. Looking at the blank dashboard, Katinka remembers how grateful she felt that her school (and hundreds of other institutions around the world) hadn't hopped on the perfotech train without vigorous debate and pilot tests. Only because the results indicated an untapped potential, did they hesitantly start monitoring their students in real time. Katinka and most of the older teachers were first opposed to yet another tool to work with in their classrooms, but when they found that the dashboard helped to pinpoint the exact student in need of support at the exact time, it became a tool they eventually leaned on daily.

The headbands – and later the lenses – have felt to Katinka like the expression of a much bigger shift in the education landscape that has taken place throughout her entire career. Take her very first class, for example. Dimitri had been a very active boy, diagnosed too late with ADHD. He'd been very bad at memorising dates, word lists and formulas. She had fought for him during the deliberations round because she felt he showed keen insights during class discussions about topics that were complex in nature. But based on his overall test scores and restless behaviour in class, the teacher panel had decided to let him repeat the year. He'd got bored, started a digital business when he was 16 and dropped out of school on his 18th birthday. She ran into him two decades later. Dimitri had a thriving business and had never regretted his decision to drop out. 'I learnt what I needed to know via online courses and forums, and by allowing myself to experiment when I was young', he said.

Had Dimitri been a student in the final year of her teaching career, Katinka was sure he would have been one of their top performers. Not in the MOOCs for universal basic knowledge that all students in the country had to take. He would have done just fine, because now the content was gamified to keep his attention, and technology such as holograms added a more immersive aspect to history, for example. But he wouldn't have risen above the average. However, the Three Pillar classes would have yielded a much better result.

He would have absolutely loved the Technology classes, where he could have experimented with digital solutions in hackathon style every week. In the labs children were allowed to move about freely, and talking – either to collaborate, or to encourage fellow classmates – was encouraged. The Creativity classes would have been a challenge for Dimitri, Katinka thought, as he had struggled to express himself. The Pure Play modules would have stumped him, but eventually he would have loosened up. The modules that opened up the senses through haptics and olfactory stimulation would certainly have worked wonders to help him feel more anchored to the present and support his personal growth. And he would have been an absolute star in her Strategy classes. Using all kinds of data about problems to theorise about solutions and then devising tests to validate proposals? Absolutely his cup of tea – he never would have been one of those students that 'forgot' his AR lenses!

A NEW SCHOOL OF THOUGHT FOR A NEW GENERATION

At the turn of the century, several issues began to rise in the field of education. The first was the fact that the amount of data children had to learn grew significantly. A person born in the 1820s had significantly less tech knowledge, formulae or history to memorise than someone living in the 21st century. The second issue was the growing rift between the theoretical knowledge learnt at school and soft skills needed to enter the job market. Lastly, children's attention span declined after the 2020s. There was a great need to move from the traditional, passive way of learning to a more active education style. Along with universal internet access, there were a handful of technologies that made this transition from passive to active education more virtual, immersive, and hands-on. These included augmented reality (AR), virtual reality (VR) and holograms. For about a decade, attention monitoring via facial recognition was used to improve the curriculums of each individual class, until they reached at least 85% of average student attentiveness.

The next two decades were filled with exciting adaptations of this concept. The first adaptation was meant to come in the form of Neuralink-like implants, but the procedure was deemed too invasive and the outcome uncertain. Instead, adaptation came in the form of attention monitoring bands that collected data via EEG tech and gave student instructions via bone conduction and sent data to their teacher's tablet. Monitoring devices were advised to be worn at home during their study-time as well. Based on the data accumulated, the device would then identify the best moments for data retention. It would then communicate with students' mobile devices to recommend them the most optimal time of the day at which to study.

As schools started to implement these digital solutions, they also gathered more data. When they finally started to pool and analyse their big data, the entire view on education began to shift away from turning children into passive knowledge repositories, and more and more towards the new model. Suddenly, 'knowing things' was only considered a starting point and not the ultimate goal. Kids were increasingly taught where to find the right information, how to judge it, how to analyse and use it. This quickly developed into the 'Base and Three Pillars' approach to education.

The idea of this new approach was to develop MOOCs (Massive Online Open Courses) that held all the basic knowledge and theory children at a certain age were supposed to acquire in a country (and in some cases where language allowed it, even across borders). These courses were fully automated to free up teacher time for the Pillars. The MOOCs were also gamified, to keep student engagement high and continuous. All children took MOOCs consecutively, only able to access the next course when the previous 'level' was fully mastered. This was what became known as the Base.

The courses in these MOOCs were enhanced with technology like holograms, so students could listen to the best performances of the best teachers in the world. Or view Hollywood-style reconstructions of historical events. What could be better than hearing Napoleon Bonaparte recounting his life and the Waterloo battle himself, rather than having to read through hundreds of pages in his biography? This meant that education centres, like schools, campuses and other learning facilities, could transform into learning labs. That's where the Three Pillars came into play: Strategy, Technology and Creativity. The first pillar, Strategy, was centred around the idea that being able to look at any kind of data and coming up with a structured, executable solution that could be verified through user testing, was a worthy skill in the job marketplace.

The second pillar, Technology, was centred around the idea that children should no longer need to answer the question 'How far?' when it came to technology, but instead 'To what use?'. Modules in this pillar were based on the sporadic hackathons that were organised at the turn of the century, where existing technology was used as a playground for solutions. In Technology classes from the 'Base and Three Pillars' approach, children were asked to create something new with any tech equipment they found in the lab. Technology skills were of course trained in these classes, but they also served to teach collaboration skills as well as the failure-as-a-learning-opportunity mindset.

The final pillar, Creativity, was aimed at keeping people's minds open. This pillar used techniques for play mindsets that would encourage out-of-the-box thinking and creative expression. In addition, they focused on growing empathy through what would come to be known as 'deep immersion'. Based on advancements made in relation to VR headsets, children were offered experiences that immersed them completely into the life and situation of someone else. One of these advancements was the dispersal of scent through little openings right below the goggles. The goggles themselves had five cartridges from where scents were mixed to fit the experience before being released. Additionally, for VR to be able to stimulate the nervous system through pressure, pain and temperature, the goggles were combined with haptic bodysuits, whose actuators placed within the fabric generated vibrations that simulated the sensation of touch. This allowed them to stimulate all five of the major senses – hearing, sight, taste, touch and smell. With all senses engaged, immersive technology could really let someone walk a mile in another person's shoes – hence 'deep immersion'.

Upon completing their studies, the end result of which we now term a diploma, was no longer based on number or letter scores. All MOOCs had to be completed, so everyone graduated with the same 'degree' around their 18th birthday.

What made their graduation special was that their achievements and expertise were officially validated. This gave them a clear view of what to specialise in if they studied further, or which type of jobs they should start looking for in the workplace.

CAN THE HEROES OF YESTERYEAR BE TOMORROW'S TEACHERS?

That attention spans are lowering with each new generation, is no secret. Researchers at the Technical University of Denmark concluded that the attention span of people all over the world is narrowing due to the amount of information that is presented to the public. The Chinese government is already implementing several measures to combat this. Notably, they're limiting the time children spend playing video games. Other measures revolve around monitoring attention with tech – what we called 'perfotech' in our proposal – like attention detecting facial recognition software, or performance monitoring hardware like bands.

Some schools and countries' education systems are already addressing problematic passive educational environments. Finland remains at the top of the rankings for the world's best education systems through the implementation of play-based learning in schools across the country. Elon Musk has also experimented with new types of education by creating Ad Astra and Astra Novia schools, where students learn through online, live team games with students from around the world. His education system, like the other advanced ones, stresses the importance of group work. Following the online learning theme, the Massive Open Online Courses (MOOCs) market is expected to grow at a compound annual growth rate of 32.8% during 2021-2026.

AR, VR, AI, hologram and other technologies are also already tapping into the field of education. With SkyView App, for example, students can explore the universe via AR overlays of the night sky. Students can point their mobile devices towards the sky to discover stars, constellations, planets, and satellites. With the Froggipedia App on the other hand, students can explore and discover the unique life cycle and intricate anatomical details of a frog.

Some medical students already have the opportunity to use Microsoft HoloLens to learn about the human body by, amongst many other things, 'flowing' through the bloodstream or 'walking' inside human body parts in order to better understand the inner functions of the body. This not only helps students gain a better understanding of anatomy but also gives them a more detailed perspective on treatment for different medical conditions.

Holograms of professors were first implemented at Tecnològico de Monterrey in undergraduate courses in 2016. They pioneered the use of holograms, captivating both students and teachers. Holograms increase the mobility of teachers, mentors, historical leaders, and others and gets them to interact with students live or via recording.

Google Expeditions were particularly useful during Covid, as teachers were able to take students around the globe, thanks to various immersive school trips offered by this VR platform. Trying to teach your students about a bombing raid on Nazi Germany during World War Two? The 1943 Berlin Blitz in 360°, an experience produced by Immersive VR Education for BBC, can help bring it to life in ways a textbook could never achieve.

There's also been an attempt to develop the world's first Multisensory Mask that releases smells, vibrates, and blasts your face with air or mist. The project, FEELREAL, has earned plenty of backers on crowdfunding sites, but hasn't provided a prototype yet.

We already got to re-live the experience of listening to Tupac, Michael Jackson, Elvis Presley and Amy Winehouse – well after their deaths – via holograms, so it's not hard to see those applied to education systems and historical figures. The closest to haptic bodysuits today is 'synthetic skin'. Developed by Northwestern University, it is a wirelessly powered, 15-centimetre square patch that can be stuck onto any part of the body, using actuators that vibrate against the skin to simulate tactile sensations.

In one of the papers published in the Soft Robotics Journal, scientists revealed details of a second skin that's about 500 nanometers thick. It was intended for VR users, and is far more sophisticated and less obtrusive than existing haptic feedback systems. The prototype offered real-time feedback at a frequency of 100Hz and a sensitivity of up to one Newton of force.

Turns out that even AR contact lenses aren't that far-fetched either. Mojo Vision is developing the world's first, true smart contact lens with AR display, powered by an external compute pack. Its hexagonal shaped display boasts 14.000 pixels per inch. Its magnification system expands the imagery and beams it into a part of the retina called the fovea. Mojo Vision's Vice President Steve Sinclair explained to Forbes that the compute pack is meant to be a device that communicates with the lens and is to be worn around a user's neck. The device is where the processor, GPU, batteries and storage will be contained.

PROBABILITY

SCIENCE

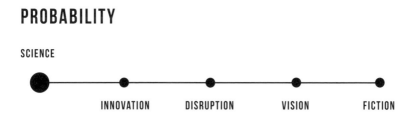

INNOVATION   DISRUPTION   VISION   FICTION

A lot of 'perfotech' mentioned in the article is already here. Their use was accelerated by Covid-19, like the Massive Open Online Courses, AR, VR incorporation and even hologram professors. Synthetic skin in the form of a patch is under development (and we're eagerly waiting for someone to make it into a full body suit). Scent-releasing VR glasses already exist, but it will take a while before they hit the market, as the FDA has not approved their usage. Another type of tech that's mentioned in our vision are the smart contact lenses. They are just an idea, waiting to be prototyped. There seems to be a lot of tech that's already in use or at the very least being developed, so why the low score?

Personalised learning is the ultimate goal of education systems around the globe. Yet it seems like moving closer towards that goal is a rather slow and costly affair that governments aren't eager to push. That cannot be said for China, where pilot projects for bands that monitor brain waves are already ongoing. In other classes, students' attention is monitored through video recording and face recognition and compared with data about their achievements. If these kinds of innovations and user testing continues, we're looking at China becoming the most dominant intellectual global force. Looking at their yearly patents growth rate, it seems that their efforts are already proving fruitful. So, are we content simply to ride their wave of success, or will we rise up and redefine our education ourselves?

GIVE A MAN A CRISPR AND HE'LL EAT FOREVER

Tom is daydreaming about augmented reality wine labels and munching on a *Blue Java* banana (that is actually blue). A beep from his app alerts him that the next check is due on his vines. His office is bathed in the warm pink light of the luminescent solar panels that cover the windows. Made from fruit and vegetable crop waste, it converts UV light into electricity even when it's cloudy (as it usually is in Glasgow). He flicks on a monitor and selects the view from one of his bumblebee drones. Wearing a tiny backpack camera, this bumblebee not only gives him the ultimate bee's eye view, it also flies around as a tiny pest controller. On its feet is an organic, inoculating fungus that it gently distributes on its foraging journey.

Through the bee's infrared sensor, the system picks up a temperature change and calls for a vine bot to take a closer look. A robot wheels over and takes a quick soil sample. The verdict: increase fertilisation. Soon a fine brown mist is sprayed over the soil – a completely odourless nutrient-rich 'worm juice' that has been rehabilitated from biosolid sludge. Tom finances most of his winemaking business through the carbon credits he earns by repurposing the waste of his apartment block. Money down the toilet? Most certainly not.

It's lunchtime and Tom decides to check on his merlot grapes and maybe enjoy a glass with his cultured meat sandwich. He steps outside his office and takes an elevator up six floors onto his rooftop vineyard. On the way, he passes the coop of dodos he is currently breeding. When companies started de-extincting animals like the woolly mammoth in the 2030s, it was only a matter of time until someone brought back the Mauritian mascot. Little did we know that its eggs are naturally low in cholesterol.

On the rooftop, Tom steps out to admire his gene-edited histamine-free chardonnay grapes alongside rows of anti-ageing merlot. As Louis Pasteur once said: 'Wine is the most healthful and most hygienic of beverages.' Turns out, he was right. Ever since scientists started tinkering with engineered yeast, they've been able to ramp up the amount of the healthful component resveratrol. This has finally led to the holy grail of winemaking: wine with no hangover. What's more, the wine club of 2050 features wines bred for specific anti-ageing benefits, heart health and fighting dementia. Bottoms up!

FARMING OF THE FUTURE

The world of 2050 is bound to be a more wholesome, bountiful place through agriculture. By then, Agriculture 4.0 will be well entrenched, transforming farming into a well-oiled manufacturing industry. Confronted with climate change, dwindling natural resources and a population explosion, governments, investors and scientists will pool their expertise to achieve the 70% increase in food we'll need. Imagine if you could bring together the best regenerative farming practices, the most intelligent technology, affordable energy and flourishing gene-edited seeds? That's the recipe for feeding the world in 2050. With the help of high-speed data connectivity via nanosatellites, smart farms will use technologies like drones, robots, temperature and moisture sensors, aerial images, and GPS technology. These precision agriculture systems – powered by endless cloud data – will make farms more profitable, efficient, safe, and environmentally friendly.

The dream of farmers remotely monitoring swarms of planter and fertiliser robots will be commonplace. Farms will become like tightly controlled factories that produce consistently reliable products. Thanks to the discovery of CRISPR-Cas9 technology, much more precise genetic manipulation will be possible. Everything will be more convenient and taste better. Sweeter fruits, less allergic reactions, no hangovers. Fruit, vegetable and livestock species will be brought back from extinction. Cattle will be bred to be more heat resistant and cope better with climate change. Crops will be tinkered with to be more nutritional, drought and pest resistant and higher yielding. Consumers will become better educated on the multiple benefits for their health and the environment. They will be more accepting of methods that, in essence, replicate mutations that take place naturally anyway.

The soil and water we will have left will be treated like gold. The world will finally realise that to keep up with a rising food demand, the answer doesn't lie in more deforestation and more megafarms but working more productively with the soil we have. With one third of earth's soil acutely degraded due to agriculture, regenerative farming will no longer be a marketing buzzword – it will be the new standard. Regenerative agriculture will shift the focus from a yield obsession towards managing a better functioning ecosystem.

The ecological wastelands created by intensive farming practices will be brought back to life by encouraging biodiversity, trapping carbon emissions and keeping soil disturbance to a minimum. Currently the experiments of smaller farm owners, it will spread to megafarms too. The biological product industry (biofertilisers, biostimulants and biological control agents) will be booming. We'll finally understand that the very health of our bodies is directly connected to the health of our soil. As the food-as-medicine trend expands and we deeper understand the link between human and soil microbiomes, consumers will start demanding that these delicate ecosystems are protected. Scientists will spend the next decades trying to understand how microbiomes in soil, people, oceans and air can be restored and protected before they go extinct. They will discover many fascinating connections between what we drink, eat, breathe and even accidentally ingest (i.e. dirt). Nothing will be taken for granted.

Even for those countries running out of fertile soil, there will be hope. If food can be cultivated on Mars, deserts can be coaxed back to life. Food can be grown in sea or saltwater. And, of course, vertical and urban farming will be huge. An aerial view of food resilient cities will show seas of green – on every possible wall and rooftop, in every possible underground tunnel. Food will be cultivated in supermarkets, malls and schools, parks and multi-story car parks. Everyone will be a farmer, growing food in a cellar, garage or ceiling. Perhaps even right there on the kitchen counter as the Personal Food Computer promised. Food will, literally, be grown from thin air – using energy efficient artificial solar-powered lighting and hydroponics.

Meat on the menu? Yes, very much so. The desire for meat is ever growing. Over the last thirty years, meat and dairy consumption has tripled in low and middle-income countries – on top of already gigantic meat appetites in developed nations. But the meat of 2050 doesn't have to come from animals. Meat taxes – already discussed in Germany, Denmark and Sweden – will curb meaty cravings. Many will follow China's lead – aiming to cut meat consumption by 50% by 2030. Soon, cattle farmers will lobby to not call this meat, and we'll have to come up with a new word like 'feat' (fake meat?) for new generations of 'cytovores' (consumers of cells).

The animals we do eat will enjoy far more respect as crucial partners in soil preservation – and more of an occasional delicacy. Most of the technology guiding human health will filter down to livestock. Health trackers/implants, breath and stool analysis, gut-specific nutrition and silvopastoral grazing (munching among shrubs and trees with edible leaves or fruits). Livestock will live lovely, coddled lives while they fertilise soil and generate 'cow power' electricity with their manure.

The result: meat that is far too flavoursome to waste on nuggets, Bolognese and burgers. For these things, there will be cellular agriculture – growing animal-based protein products from cells, not animals. By 2022, more than 40 companies will already produce bioreactor-grown meat so we can expect a massive influx of meat breweries creating 'craft' meats.

AGRITECH GOES MAINSTREAM

Big robots, small robots, sensors, tags, drones, beacons – we know future farms will have a lot of them. And they will all communicate with one another, log data on the cloud and report back to a central app. While the farmer sleeps, sensors measure interactions between plants, soils, and animals. Robots, AI and algorithms will work on new food sustainability practices and models. They will creep and fly around your vineyards, fields, tunnels and plant walls, weeding, monitoring, picking, hoeing, seeding and more.

Today, a swarm of six Fendt Xaver robots can cover around three hectares per hour – each three-wheeler the size of a souped-up motorbike. Everything about these smart planters makes sense: being battery operated, they emit less CO_2, are less noisy and safer to operate than giant machines. Less soil damage, less oil spills, less heat, less power, less pesticides. These robots are green energy ready, able to be powered by a biogas plant or photovoltaic facilities, wind power or fuel cells. And, of course, they can work 24/7 and dramatically increase yield. They can vary seed depth and pressure based on real-time sensing of soil moisture, temperature, humus content and plant residues. All a farmer must do is park the docking station near the field where the bots will operate, fill them up with seed and hit 'go' on the app. They automatically go to work, and when they need a charge or seed top-up, they drive themselves back to the charging station. If one stops or jams for whatever reason, another seamlessly takes its place.

In a bid to address the seasonal worker shortage, scientists and mechatronics experts are hard at work to bring farmers fruit picking solutions. These will become a common assistance during harvest time when you could have humans picking during the day and robots at night. Octinion's autonomous strawberry picking robot Rubion can gently pick berries just like a human picker, without bruising the strawberries. Thanks to built-in quality monitoring, the robotic system allows for sorting, advanced crop monitoring and precision farming. Over in Australia, Monash University researchers have developed a robot capable of performing autonomous apple harvesting. Using cameras to scan the trees, the robot can harvest an apple in around seven seconds. Compared to the four or five seconds it takes a human to pick an apple, this is already very promising. Naio Technologies have three farm friends to choose from: Ted the vineyard weeder, Dino the vegetable weeder, and Oz the...well Oz can do almost anything from weeding and hoeing to seeding and transporting.

Tevel's FARs (flying autonomous robots) are flying fruit pickers that combine cutting-edge algorithms, AI, and data analytics. They can spot and pick ripe fruit from the sky, and work 24/7. While these types of cobots (collaborative robots) always bring up the question of taking jobs from humans, Tevel believes that a worker shortage will see a useful collaboration between man and machine. What's more, just because fruit picking is a centuries-old trade doesn't mean this low paid, repetitive work should exist forever. The World Economic Forum estimates that automation will create 97 million jobs by 2025, more than it will displace. Cobots will augment and enhance human strengths with the precision and data capabilities they bring.

However, flying assistants are not just for affluent countries. In sub-Saharan Africa, the not-for-profit organisation TechnoServe uses remote sensing, drone mapping, machine learning, and satellite data to help boost cashew nut production in the West African country of Benin. This pilot project has been instrumental in helping smallholder farmers know where best to plant their trees and increase the quality and quantity of their yields. These 'eyes in the sky' will be game changers for African farmers, helping with everything from land registration and locating livestock to taking inventory of crops and estimating crop yields.

It's not just crops getting tech upgrades. GEA's DairyRobot is more than just a robotic milking system – it delivers a precise image of every udder quarter of every cow to quickly pick up differences in milk quality or changes in temperature (e.g. detecting udder infection). Ceres Tag's solar-powered, satellite-connected smart tags constantly transmit real-time data back to the farmer. They monitor everything from a cow's location and feed intake to its general health and fitness. MooCall's smartphone-connected sensors fit onto the tail of a pregnant cow, then send the farmer a text message when the animal is approaching calving. MooCall's Heat collars and ear tags monitor mounting behaviour and bull activity levels to determine – pretty accurately – when a cow or a heifer is in heat.

One day, these smart collars and livestock wearables will monitor absolutely everything – just as they do with humans right now. Imagine Bluetooth-enabled sweat strips that measure sodium, potassium, and glucose levels. Imagine using the same technology as human breath biopsies to analyse a pig's breath for nutritional problems. With a smartphone, a farmer will have many apps for on-the-spot diagnoses such as detecting metabolic diseases in cows and pigs from just a few snapshots.

EVERYTHING IS FARMLAND

In 2050, what we know as farms – big swathes of land on the outskirts of cities – will be something completely different. In fact, in a few decades, every-one will be a farmer. It took a pandemic to boost a huge interest in gardening and urban farming as everyone wants to try their hand at cultivating their own food. From balconies to cellars, from patios to attics, food is being grown without sunlight or soil, with sophisticated growing kits.

Singapore is a promising test case for modern day urban farming. A country that imports 90% of its food, it has set itself the challenge to produce 30% of its own by 2030. With virtually no land to speak of, it's creating hydroponic farms on parking structure roofs (Citiponics), retro fitting vertical farms into existing buildings (Sustenir Agriculture) and converting unused spaces into greenhouses. Agritech business AbyFarm is leading the farming revolution, with the aim to transform existing high-rise buildings, car parks, rooftops and land into automated, smart remote-control farms. AbyFarm uses intelligent, AI cloud-driven machinery – IoT, sensors, machinery, blockchain, data analysis, high-tech self-regulating farming system – to attain high yield, sustainability, and traceability.

The World Wildlife Fund is also throwing its weight behind research into new ways to lower the massive environmental footprint of growing food. One such project is trying to ignite a vertical farming revolution by bringing together different stakeholders in St. Louis, Missouri. Here, farms are created in naturally cool areas such as a network of underground caves once used to brew beer, cold storage in postal hubs, or in spaces next to power plants that can take the excess heat from an indoor farm and convert that into energy.

In London, Growing Underground grows micro greens and salad leaves 33 metres below the busy streets of Clapham. Here, in this entirely artificial environment – World War Two bomb shelters in a previous life – they grow food in just two weeks to have it on your plate in four hours – a locavore's dream. With the help of hydroponic systems and cutting-edge, LED technology, crops are grown year-round in the perfect, pesticide-free environment using 100% renewable energy. Everything – from temperature to illumination – is monitored closely by sensors, with data sent directly to Cambridge University's engineering department to work out new routines for future crops. Coriander, for example, can now be grown in 14 instead of 21 days, and research suggests that this accelerated growth could work for carrots and radishes too. While this may not overhaul the potato industry overnight, these precise controls of indoor farming spell an exciting future for agriculture.

VERTICAL FARMING – THE SKY'S THE LIMIT

With massive innovation in lighting technology over the years, vertical farms have gone from science projects to proper legitimate businesses. Aerofarm, Bowery, Gotham Farms and Plenty are some of the names to watch. These businesses have big goals and celeb funding to match. Robert Downey Jr. is funding Ÿnsect, while Bowery has attracted investment from Natalie Portman and Lewis Hamilton.

Highly efficient, indoor vertical farming is what German start-up Infarm excels in. Their innovative modular farms can be found in over 30 cities in the world – including more than 1300 supermarket aisles (e.g. Marks & Spencer in the United Kingdom and Kroger in the United States). In 2021, the company released new, high-capacity, cloud-connected Growing Centres. Its vision: a globally interconnected, sustainable and highly efficient farming ecosystem, with 100 Growing Centres by 2025. Each centre is designed to function both as a local farm and a distribution centre. It comprises dozens of modular farming units (standing 10-18 metres high) and occupying a 25 square metre ground footprint. Taking just six weeks to build, it promises to yield the crop-equivalent of 10,000 square metres of farmland. The centres work on the same distributed approach as their in-store equivalents. They use a combination of big data, IoT, and cloud analytics to measure and deliver the precise energy, water and nutrition to individual units, in order to maximise yield and minimise resources used. The entire Infarm network is connected to a 'central farming brain' that gathers more than 50,000 growth, colour and spectral data points.

It makes sense for supermarkets to get in on the vertical farm action. Responding to the fact that fresh food in Europe travels on average around 1000 kilometres from farm to shelf, German retailer REWE wants to reduce this to a few metres. Its Wiesbaden Market concept store features a rooftop farm that will, for example, produce around 10,000 kilograms of fresh fish and 800,000 pots of basil. Growing in harmony together, the basil will filter the water while the fish faeces fertilise the plants.

Online grocer Ocado has invested in one of the world's largest vertical farms by JFC – dubbed 'The Garden of England'. It will be the size of 70 tennis courts and be able to supply more than 1,000 tonnes of fresh produce to thousands of UK supermarkets, when fully operational. It has also formed a partnership called Infinite Acres with Netherlands-based Priva Holding BV, a horticultural technology provider, and indoor farming experts, 80 Acres Farms.

Don't think vertical farming will only dish up kale and strawberries. Ÿnsect uses vertical farming to breed mealworms, which are then transformed into food for fish, plants, pets and eventually humans. This has important potential for farmed fish, since trawling the ocean floor for anchovies isn't sustainable. Ÿnsect's mealworms can provide fish with a lower-cost, higher-quality protein. In Singapore, it will only be a matter of time before the Apollo Aquaculture Group has one of the world's largest vertical fish farms up and running – a whopping eight stories high. It won't just be the highest fish farm – it will also produce an amount of fish and shrimp that is six times higher than established aquaculture operations in Singapore (measured in fish per ton of water). And finally, Infarm is already developing a breeding programme to adapt grains, legumes and other staple crops for vertical farming. Because, after all, with CRISPR-Cas9, anything is possible.

OPENING OUR MINDS TO GENE EDITING

Gene editing is another hot topic for the future of food. If there was a way to grow plants and livestock that was naturally more resistant to disease, and crops that were higher-yielding, why wouldn't we pursue it? This is what the NextGen Cassava Project is hoping to achieve. Researchers from this pan-African group are studying cassava genes in the hope to fix its vulnerability to the mosaic virus, plus to improve yield and make it more nutritional. While there is often more focus on grains like wheat and rice, for this African staple to have a higher starch content would be a game changer.

If we can remove pits from cherries, keep mushrooms from browning (already done) and make sweet-tasting leafy greens, why not? With the discovery of CRISPR-Cas9 technology, scientists can now delete or multiply certain genes to stop them producing certain proteins, or to make more of them. Think of a DNA as a string of letters that make up a genetic code. Once we find out what each letter does and how it affects a plant's performance, letters can be deleted or changed. Gene editing is basically replicating this process, instead of waiting for it to happen. People tend to be fearful of gene mutations, yet plants – even us humans – mutate all the time.

Crop strains created by gene editing are already coming to market, for example, Artesian by Syngenta, and AQUAmax by DuPont. Low or no gluten wheat could be on the cards, nuts that don't cause allergies, caffeine-free coffee, soybeans that are lower in unhealthy saturated fatty acids, and flax seeds with the same high omega-3 fatty-acid content found in fish. Japan's heart-healthy, super-tomato might be a Frankenfood for some, yet this is only the beginning. Other genome editing projects in Japan include efforts to breed meatier sea bream, hypoallergenic eggs and higher-yield rice plants.

There are projects under way to breed cattle with less hair, allowing them to sweat more and cool down. Gene editing has been successfully used to produce pigs resistant to the fatal porcine reproductive and respiratory syndrome. Start-ups like Pairwise are focussing on removing pits from cherries and creating sweet-tasting leafy greens.

And what if you could actually count your chickens before they hatch? When breeding layer chickens, the only way to differentiate male from female is after birth. Israeli company EggsXyt used CRISPR to transfer the DNA of a jellyfish into the male chromosome of laying chickens. Working like an egg ultrasound, when a bright light is shone through the genetically modified eggs, the female eggs will show nothing, but the male eggs will appear blue. *Voila!* No more killing four billion day-old male chicks a year, and instead, adding four billion eggs (and extra profits) to the food supply chain.

HEALTHY SOIL = HEALTHY PLANET

Carbon offset projects come in many guises but there will be a continued focus on the role of soil in bettering our climate. It's simple: healthy soil gives us healthy plants, and nutritious plants give us healthy bodies. Moving towards regenerative farming is the best way to encourage carbon drawdown and maintain healthy soil from turning into dirt. When we speak of healthy soil, this is an intricate web of fungi, bacteria, nematodes, mites, termites, and earthworms that all work together for properly functioning soil. Microbes fix nitrogen from the air into soluble nitrates that act as natural fertilisers; a deeper understanding of them will be an exciting development in agricultural biotechnology.

For example, BioEnsure by Adaptive Symbiotic Technologies is a fungal seed and plant treatment that, when sprayed onto seeds, helps plants to adapt to water-related stress. Companies like Koppert have pioneered the use of natural pest control and bumblebees for natural pollination and develop microbial products that strengthen and protect crops both above and underground. In the future, we can finally move away from harmful pesticides as we find the bacteria, fungi, viruses and yeast cells who can safely do the job just as well. Farmers will actually start to earn an income from feeding their soil to sequester carbon. Take the *Carbon by Indigo X Corteva* example. Seeds and farm chemicals company Corteva has created The Corteva Carbon Initiative, which pays farmers around $15 an acre to shift to practices that pollute less, use fewer chemicals or farm crops that pull carbon from the atmosphere and lock it in the soil. All this is tracked by the *Carbon by Indigo* app. If farmers tick the right boxes, the app generates registry-issued carbon credits, which are increasingly in demand by major corporations. Soil as a bank has great potential.

PLANTING THE UNPLANTABLE

The Loess Plateau in western China is a great success story of how a desert was rehabilitated back to the fertile land it was 3000 years ago. Googling 'Loess before and after' makes one gasp at what is possible if a region the size of France can be transformed in as little as 10 years. This is exactly the kind of big thinking Dutch firm Weather Makers are doing. Their ambitious 'Greening the Sinai' project wants to combine data from multiple disciplines to bring back the 'Garden of Eden' in as little as 20 to 40 years. If successful, it can add more moisture to the region, and can even positively influence the larger weather systems that cause extreme weather around the Mediterranean and the Indian Ocean. It's all-hands-on-deck as scientists are going to restore a lagoon – Lake Bardawil, then the surrounding wetlands, eventually bringing biodiversity to the area.

Africa's Great Green Wall is another ambitious greening project, with the goal of restoring 100 million hectares of currently degraded land, sequestering 250 million tonnes of carbon and creating 10 million jobs in rural areas. It's about 15% underway already, and will provide food security, jobs and hope for the millions who live along its path. When it's completed, the wall will be the largest living structure on the planet, three times the size of the Great Barrier Reef.

From deserts and marshlands to salt lakes, the world's brightest minds are greening wastelands, reforming weather patterns and creating green oases out of nothing (even on Mars!). Companies like Red Sea Farms are growing food in greenhouses cooled by saltwater and are working on technology that can block infrared light to prevent overheating. In a country like Saudi Arabia, where most of its food is imported due to scarce water resources – or other remote island nations – this offers great hope.

In the Praia Seca region of Brazil, the world's largest hypersaline lake, bio-technologist Camila Reveles is experimenting with growing salt-tolerant crops – specifically salt-tolerant salicornia (sea beans). Canadian start-up Agrisea is collaborating with rice farmers in the Mekong Delta to develop strains of salt-loving rice. They are using CRISPR to insert a DNA sequence into the rice that turns on genes that enable the plants to thrive in saline environments. Who knows? Very soon, green beans could be persuaded to grow like sea beans in saltwater, or asparagus like samphire.

THE END OF ANONYMOUS FOOD

Knowing where your food comes from is a trend that will continue to grow and become a competitive advantage – especially with younger demographics like Generation Alpha and Beta (born post-2010). Growing up in a super-connected world, accustomed to on-demand information, these buyers will finally bring an end to the feverish consumerism that's eating away at our planet. Instead of chasing 'bigger, faster, higher', they'll use their spending power on a lifestyle with positive social and environmental benefits. They will question every aspect of food production and demand complete transparency. This endless curiosity will be fuelled by a simultaneous explosion in technology. The blockchain will allow for complete tracking and tracing of the entire food supply chain. From scanning the barcode of your bean burger, you'll be able to see where every single bean, grain and kernel was farmed, whether the farm used herbicides or bio pest control, the status of their carbon bank, the renewable energy used, the factory where it was made and what assembly robots they use, expiration dates, storage temperatures, shipping details, and much more.

Hedera Hashgraph, the third generation blockchain technology backed by companies like Google, IBM and Boeing, provides a trusted way to track the agriculture supply chain. One example is that of Entrust, an Australian supply chain platform which can secure the integrity of a bottle of wine from grape to glass to help combat the three billion dollars in annual wine fraud. Then there is also Brazilian company Agryo, who wants to ensure that small and medium-sized producers have the same access to credit as their larger counterparts. Its creditworthiness prediction model integrates producers' financial and management data with satellite, climate and agronomic data to supply producers with a cost-effective way to get a credit score. This score is then recorded by Hedera so that its authenticity and creation date can be confirmed when used by the producer to obtain credit.

Having a decentralised, undisputable, tamper-proof ledger of transactions will be useful for most agricultural value chains, from tracking the temperature of export fruit to tracing meat parts. Hedera also has all the right sustainability chops. The network uses only a fraction of the electricity of the previous generation of blockchains. It can process more than 10 000 transactions per second, reach transaction finality in three to five seconds, and costs are limited to $0,0001 per transaction.

PROBABILITY

SCIENCE

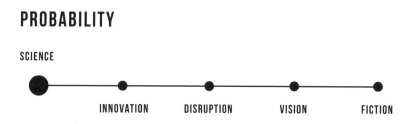

INNOVATION   DISRUPTION   VISION   FICTION

Get it right, and we'll eat like kings. Things are going to start coming to-gether for farmers, food and the planet. Agriculture has been an ageing industry for far too long and will need to embrace a digital, connectivity-fueled transformation to survive. We'll see lots of new agronomy jobs pop up, like agtech engineers, carbon bankers, soil rehabilitators, swarm captains, virtual herders and agri-geneticists.

This is where inventions will abound, and history will be made. From field to fork, we'll learn to respect not only the food we eat but the people who make them. Farming is the biggest, most important job on earth, and we can't wait to see its revolution.

ENERGY TOO
CHEAP TO METER

March, 2051, 5 am. *It's a sweltering hot day in Durban – the mercury rising just above 39°C. The 75% humidity doesn't help either. Mlungisi grabs his last kombucha from the fridge to cool him down before the big day ahead. As he closes the fridge door, it flashes an order confirmation sent to his local giga-market to restock his favourite drink. He has thirty minutes before he needs to head off to the stadium to enjoy a breather and plan his day. He lies back on his smart sofa that automatically moulds to his body shape according to its pre-programmed memory, enjoying the crisp air of the air conditioner blowing over the room. As he checks emails and schedules on his smart glasses, he hears a beep from the other room. It must be his SolarHydro hydrogen battery that has run low on stored power, clicking back into the solar panels on his roof to pull in new energy. Mlungisi happily pays his flat-fee subscription to SolarHydro which affords him unlimited electricity just like his unlimited 5G internet plan.*

Today is day three of the Commonhealth Games, held in Durban for the first time. Bringing together like-minded nations that have pioneered renewable energy, it includes a range of reimagined summer and winter sports in simulated environments. Mlungisi is the image engineer of all the South African teams. Today will see the surfing team compete against the USA, China and Japan. The Springboks are hailed as the favourites to win the indoor surfing competition.

He hops on his suave Batman-black enclosed electric tricycle, starts the engine with a fingerprint scan and sets off to the stadium. It's a hive of activity since the government's ambitious port revamp was finally completed in 2049. It now boasts colossal container storage capability and it can accommodate much larger vessels bringing spectators from abroad.

The jewel in the crown? The zero-emission natural gas power plant that now produces enough blue hydrogen to power the city of Durban and enough green ammonia to export to Europe. It has become an important refuelling stop for green ammonia cargo ships too. Run 90% on renewable energy, this eco-system has been a much-celebrated success story of the country's establishment of a hydrogen-linked 'Platinum Valley' in the 2020s. This successful industrial corridor pulls from SA's massive platinum reserves (the largest in the world) starting in the province of Limpopo, and carries through to the Johannesburg-to-Durban corridor. It's powered by one of the country's largest solar farms located in the KwaZulu-Natal Midlands.

As Mlungisi pulls into the stadium, he can't help but gasp at its magnificence. Inside, it's basically like four seasons under one giant dome. There is the water world with soliton wave technology for today's surfing competition (inspired by Kelly's Slater pioneering Surf Ranch). Next week will see an extreme kayaking race take place. Powered by hydrogen-fuel hydrofoils, it can serve up waves of up to eight feet. Next door, there is a multi-purpose ice rink cum skate park, where the first ice rugby game will kick off later this afternoon, followed by a skateboard competition between humans and bipodal robots. Back to ice tomorrow, teams will face the ultimate Yukigassen (snowball fight) challenge.

Before meeting his teams, Mlungisi spots the RainDance powership pulling out of the harbour. This floating powerplant-meets-desalination-hub is heading to Port Nolloth on the far north-western coast of South Africa to deliver water to an area plagued by drought. It makes him smile to think that, as the ship sails, it is filling up reservoirs of fresh drinking water that will quench the thirst of humans and animals living 1700kms away. The wonders of technology never cease to amaze.

COMPLETING THE ENERGY PUZZLE

Imagine living in a world with abundant green energy and unlimited fresh water for everyone. Sounds utopian? Not if we get the mix of affordable, clean energy right. The famous vision of Lewis Strauss, chairman of the US Atomic Energy Commission (AEC), came from a 1954 speech where he predicted that 'it is not too much to expect that our children will enjoy in their homes electrical energy too cheap to meter'.

Nearly 70 years on, we've still not seen this come true. The only chance we have to meet the world's energy and climate demands is if nuclear energy becomes a bigger part of the energy-generating mix. The time is ripe for a complete nuclear rethink. Today, most commercial nuclear plants are still run on uranium – a fuel that will not see us through to 2051 (not to mention the conundrum of what we're going to do with the nearly 300,000 tons of spent nuclear fuel piling up at reactors around the world). Alternatives like thorium are three times more bountiful in the Earth's crust than uranium, produce much less waste compared to other nuclear fuels, and the radioactivity levels of thorium waste fall in a much shorter time period. Thorium is fertile rather than fissile, so it can only be used as a fuel in conjunction with a fissile material such as recycled plutonium. Thorium fuels can breed fissile uranium-233 to be used in various kinds of nuclear reactors, like heavy water reactors, molten salt reactors, high-temperature gas-cooled reactors, light water reactors and fast neutron reactors.

Thorium was already under the spotlight by Oak Ridge National Lab in the 1960s. Under the leadership of director Alvin Weinberg, the famous Molten Salt Reactor Experiment (MSRE) operated more than 13,000 hours during its four-year run. The MSRE proved that a fission reaction in molten fluoride salts could be contained in Hastelloy-N, and that a molten salt-fuelled reactor concept was viable. The reactor was fundamentally different from all others: instead of fuel sitting in the reactor core while coolants circulated through, the molten salts act both as a carrier and coolant for the fuel.

The idea of building a molten salt reactor dates back to 1946, when the US Air Force created plans for a nuclear-powered supersonic jet. Weinberg had a vision for the future that would use liquid and thorium-fuelled reactors to make electricity, plus turn ocean water into fresh water. He wanted to go beyond the constraints of fossil fuels, hydropower and existing nuclear technology. Get it right and you'd not only boost USA power supplies but also aid parts of the world with little access to reliable power and fresh water.

At Oak Ridge, Weinberg and his team used molten salt technology to develop high-temperature, low-pressure, passively safe reactors. They had all the makings to work as breeder reactors. It could make fuel as it operated, with no need for solid-fuel changeouts and fuel- and control-rod mechanisms. Yes, there were concerns about corrosion but given the time, they would've come to the safety controls that exist today. Alas, the project lost funding and was shut down in 1973. The West chose to back uranium. Why? Because nuclear energy and atomic bombs have always gone hand in hand. Uranium's by-products are much easier to weaponise. The investment in solid fuel was too far along to change.

The molten salt concept lay dormant for many decades, until the year 2000, when NASA engineer Kirk Sorensen came across a book describing what the programme had accomplished at Oak Ridge. He tracked down the old technical reports, begged NASA to pay to have them scanned and uploaded it to a website he funded himself (energyfromthorium.com). As more and more people became interested in this new (old) idea, the concept slowly gained traction. Sorensen founded one of the first molten salt start-up companies in 2011. Called Flide, it believes that the Lithium Fluoride Thorium Reactor (LFTR) is the key to producing lifesaving cancer treatments and clean, reliable, sustainable energy. Sorensen has been on a mission since to commercialise the LFTR that was envisioned at Oak Ridge back in the 1960s.

According to Flide, LFTR will be the most efficient energy source ever developed. Because a fission reaction releases millions of times more energy than a chemical reaction, a liquid-fuelled reactor can take advantage of this efficiency. Solid-fuelled reactors only use one percent of their fissile material, discarding the rest as nuclear waste. On the other hand, a liquid-fuelled reactor can consume almost 100% of the fissile material. That means no long-lived radioactive waste and the added bonus of increased fuel efficiency. LFTRs also produce no CO_2 or other emissions harmful to the environment.

Who is first in line actually testing a LFTR? China, of course. Like Sorensen, it launched its molten-salt reactor programme in 2011, investing $500 million into the project. In 2021, it will switch on its experimental thorium reactor located in Wuwei on the outskirts of the Gobi Desert. It will be the first molten-salt reactor operating since Oak Ridge in 1969. Starting quite small, it will produce just two megawatts of thermal energy – enough to power up to 1000 homes. If this 'perfect technology' is successful, it has powerful potential. China hopes to build a 373-megawatt reactor by 2030, which could power hundreds of thousands of homes.

Other countries, including the United States, United Kingdom and Germany, are also testing thorium as a fuel and are working on molten-salt reactors to generate cheaper electricity from uranium or to transform waste plutonium as fuel. As China makes headway towards carbon neutrality, it may very soon replace coal-fuelled power plants with new types of reactors, or retrofit existing ones. The Chinese government plans to build more of these reactors across the deserts and plains of western China, as well as in up to thirty countries.

As a very power-hungry nation, India is also pouring huge investment into nuclear programmes. With dwindling uranium deposits and one of the largest reserves of thorium in the world (especially in its beach sands), it sees thorium as a long-term hedge. Although India is fourth in the world for installed wind capacity and fifth for solar, that won't be enough to guarantee carbon-free energy for a population that could be well over 1.7 billion by 2060. Its only saving grace will be nuclear energy.

MINIATURE POWER STATIONS

When most people think of nuclear power stations, they probably imagine something like Springfield in the *Simpsons* – vast towers with billowing smoke, bubbling liquids and luminous radioactive rats. But the future of nuclear is smaller, much smaller. China's prototype measures three meters tall and 2.5 metres wide – about the size of a small caravan.

Instead of two or three enormous power stations, several companies are pioneering smaller designs. Small modular reactors (SMRs) are nuclear fission reactors that are a fraction of the size of conventional reactors, making them much more affordable. Parts can be manufactured in factories and shipped by truck, rail or water for on-site assembly, making the process much faster and safer.

This type of technology is nothing new. SMRs were first developed in the 1950s for use in nuclear-powered submarines and icebreakers. But they have been virtually non-existent in power generation up to now. Many countries are working on these safer, cheaper, less wasteful small-scale reactors. Russia was the first to put an SMR live. Called Akademik Lomonosov, it's also the world's first floating nuclear reactor. It is expected to generate enough power to serve about 200,000 people and have a lifespan of 40 years. Danish start-up Seaborg has also come up with turnkey floating power plants that are completely modular and can produce from 200 to 800 MW of electricity (with a lifespan of 24 years). Whereas MSRE used liquid fluoride salts, Seaborg is proposing to use another molten salt, namely sodium hydroxide (NaOH). Their novel reactor concept is made of metal alloy tubes carrying flowing molten fuel salt, which pass inside a larger tube of molten salt working as a moderator. Seaborg is proposing to mount its reactors on floating nuclear power barges and supply power to shore as a direct replacement for coal power plants, starting in southeast Asia.

France is pouring €1 billion funding into state-backed utility EDF to help it develop its own SMR technology by the early 2030s. In the UK, a consortium led by Rolls Royce will develop a 440MW SMR, with plans to construct up to 16 SMRs with government funding. Oregon-based NuScale was the first company to get approval from the US Nuclear Regulatory Commission to build test SMRs in Idaho.

Just think what we could do if everyone in the world had access to a nuclear reactor blueprint? That's exactly how the OPEN100 project wants to share its vision for 'cheap nuclear'. It will provide open-source blueprints for the design, construction, and financing of a 100-megawatt nuclear reactor, which can be built for $300 million in less than two years. This will significantly decrease the per-kilowatt cost of nuclear power. Artist impressions of these (relatively) tiny plants show them in city centres, surrounded by greenery. While many regulatory mountains will have to be moved to make this happen, it's exciting to think that these smaller powerhouses could safely live among us, generating endless cheap and green energy.

With nuclear energy, there is always the elephant in the room of nuclear waste. SMRs can be built on the sites of retiring coal or nuclear plants, making them more acceptable to local communities. Oklo, a new start-up, is working on fast reactors that use the spent fuel from conventional nuclear reactors to operate. These modern-looking A-frame structures look nothing like the massive, unsightly towers mentioned before. Oklo's first reactor will be one of the world's smallest, producing just 1.5 megawatts of electrical energy. The vision is that they would seamlessly slot into city design, powering factories, campuses, large businesses or very remote locations.

One thing is certain: the SMR industry is going to boom over the next few years. According to the International Atomic Energy Agency, there are currently almost 70 different SMR technologies under development.

UNLOCKING THE HYDROGEN ECONOMY

Another much-hyped and potentially game changing energy carrier is hydrogen. According to the International Energy Agency (IEA), demand for hydrogen is expected to grow eight times to satisfy an ask of over 550 million tons in 2050. This will be as a feedstock, but also for transportation, building heat, and power generation. The hydrogen that is currently produced – known as 'grey' hydrogen – comes from natural gas and generates significant carbon emissions. The cleaner version of this is 'blue' hydrogen, where carbon emissions are captured and stored, or reused. The ultimate version? 'Green' hydrogen, which is generated by renewable energy sources (sun, wind, tides, hydro, biofuels) without producing carbon emissions in the first place. While this idea is exciting, it's currently very expensive to produce.

Green hydrogen is somewhat of a chicken and egg situation. Solve the cheap electricity conundrum, and you'll have green fuel for days. Hydrogen is produced through a process called electrolysis where water is split into its components: oxygen and hydrogen. This hydrogen can be transported across any distance, either in liquid or gaseous forms, or mixed with other elements like ammonia or methanol. In the short-term, blue hydrogen will be cheaper than green hydrogen. But this will change fast as the costs of electrolysers and renewable power will come down fast. BloombergNEF predicts that green hydrogen will outcompete blue hydrogen on cost by 2030. But that's not to say it's remotely affordable yet. Bank of America analysts estimate that green hydrogen prices would need to fall by 85% to be competitive with regular hydrogen – something only likely to happen by 2030.

If you want one more colour of hydrogen, it's 'gold' hydrogen. This natural hydrogen in the Earth's crust will soon cause a gold rush. Just as oil barons drilled into the ground in search of fuel, hydrogen prospectors will chase fairy circles to try to work out where this gas can be found. Earlier, we mentioned electricity too cheap to meter. How about electricity that's completely free? That's what the African village of Bourakébougou, 60km from Mali's capital of Bamako, is enjoying. Here, natural hydrogen wells are used to produce clean electricity, distributed free of charge to the local population. The company in charge, Hydroma, came across the H_2 by chance, whilst drilling for fresh water. Instead, it found the purest naturally occurring hydrogen ever discovered: estimated at 8km in diameter with a concentration of 98%. Imagine sitting on that sort of energy goldmine. Over the time, the hope is to not only meet Mali's energy needs but also other countries on the African continent. And Mali might not be a once-off. There are reports of hundreds of global occurrences of over 10% gas concentration, which could mean a potentially inexhaustible source of green energy.

While green hydrogen is first prize, we need governments to support both electrolysis and carbon capture hydrogen. As with nuclear reactors, once the right policies are in place to incentivise or make green energy mandatory, countries can make the switch from grey to green hydrogen and change from coal and gas to hydrogen. The dominoes just need to start falling... One of the most promising and groundbreaking carbon capture technologies is the Allam-Fetvedt Cycle – essentially a zero-emission natural gas power plant with hydrogen production.

This cycle uses the oxy-combustion of carbon fuels and a high-pressure supercritical CO_2 working fluid in a highly recuperated cycle that captures all emissions by design. The only by-products: liquid water and a stream of high-purity, pipeline-ready CO_2. The beauty of this process is that it can use many types of fuel, from natural gas, unprocessed raw and sour gas, to gasified solid fuels such as coal or biomass. The 50 MW Allam-Fetvedt Cycle demonstration facility is currently operating in La Porte, Texas, with several natural gas projects currently in development. They've also made good progress on a coal-based system.

HOME POWER PLANTS AND HYDROGEN CARS

Will we see hydrogen power in our homes? Most definitely. In 30 years' time, there will be no excuse not to have your home off the grid, as solar panels will be as affordable as painting your roof white (which we should all do). Tesla's Powerwall uses lithium batteries as a way to store the energy you generate. But soon, you could also integrate a hybrid hydrogen battery to keep the lights on. Australian energy company Lavo has created a storage system that connects to a home's solar inverter and mains water, through a water purifier. Using solar energy to electrolyse the water, it splits oxygen and hydrogen. The oxygen is released and the hydrogen is stored in the LAVO's patented metal hydride 'sponge'. The hydrogen gas is then converted back into electricity when it is needed, using a fuel cell. Voila – your very own small power plant. Lavo boasts almost three times the capacity of Powerwall 2 and stores enough to power a home for two days.

Apart from clean electricity, green or natural hydrogen can also be used to decarbonise the transport sector. Japan is one country that doesn't want to be dependent on fuel imports forever. As a pioneer in hydrogen technology and a champion for carbon capture and storage, Japan sees hydrogen as the fuel of choice in its quest to reduce emissions from all sectors. Automotive heavyweights, Honda and Toyota, are leading the way when it comes to fuel cell technology development. In case you need a refresher, fuel cell electric cars beat battery-operated ones since they can generate power through a chemical process using hydrogen fuel. Thus, there is no extra pressure on the electricity grid to charge batteries. Can you imagine if every household charges their car at the same time at night? Complete chaos.

Toyota is coming to the party by dramatically reducing the price and size of the system compared to previous fuel cell vehicles. By 2050, Toyota aims to cut global average CO_2 emissions from its new vehicles by at least 90% compared to 2010. The manufacturer has also trialled fuel cell technology for forklifts, buses and delivery trucks. Honda is using electricturbo air compressors for a pioneering system that generates the necessary electricity for propulsion by using an enhanced hydrogen and air mixture.

In the USA, California has the largest number of hydrogen fuel cell electric vehicles (FCEVs) of any state and one of the largest hydrogen refuelling station networks in the world. It will be interesting to see how its hydrogen fuel journey unfolds as more and more manufacturers make and sell hydrogen fuel cell vehicles. We're seeing much more variety in models and prices have already come down. Not stopping at cars, the California Fuel Cell Partnership has a vision of 70 000 heavy-duty fuel cell electric trucks on the road by

2035, supported by 200 hydrogen stations. That's a bold goal, but one way to get to 100% zero-emission trucks by 2045.

We know this transition is going to be gradual and that fossil fuels won't disappear overnight. Porsche is one car manufacturer that believes that there can be a place for internal combustible engines in the world – it all just comes down to clever engineering of carbon emissions. Teaming up with Siemens, Porsche is building its first synthetic fuel production plant in Chile, hoping to produce its first batch of carbon-neutral fuel in 2022. Initially, this will be used for its racing cars and Experience Centre cars, with the hope to produce 550 million litres by 2026 and being completely carbon neutral by 2030. The plant will generate clean electricity from wind turbines built by Siemens, then make fuel by dissociating hydrogen and oxygen molecules from water. CO_2 filtered from the air will then be combined with hydrogen to make synthetic fuel.

KEEPING OUR COOL

We need energy to switch our lights on, cook our food, run our businesses, get from A to B. But how often do we talk about the energy it takes to keep us comfortable? Air conditioning is what the IEC calls one of the 'blind spots of the global energy system' – an industry that has evolved very little over the last 100 years. The world faces a 'cold crunch' if, by 2050, two-thirds of the world's households could have air conditioners.

Things need to change, fast. Air conditioning is one of the biggest culprits for CO_2 emissions. Globally, about 12% of non-carbon dioxide emissions come from refrigeration and air conditioners, according to the US Environmental Protection Agency. And then there is a humungous amount of energy they gobble up. As the middle class grows over the next few decades, global energy demand from air conditioners is expected to triple by 2050, requiring new electricity capacity the equivalent to the combined electricity capacity of the United States, the EU and Japan today.

Air conditioning perpetuates a vicious cycle: the hotter it gets, the more we blast cool air. The more we use air conditioning, the warmer it gets.

For countries like India and Bangladesh, it's not a question of comfort but of survival. Research has shown that some areas of northeast India will become so hot that being outside for more than a few hours could be deadly.

So what's the solution? For start-up Gradient, it's by creating HVAC systems that use very little energy to heat and cool, and to power them with renewable energy. Gradient is solving the first ask by using a different refrigerant (R32) to the usual hydrofluorocarbon refrigerants, known to be over a thousand times more potent than carbon dioxide. Gradient's heat pump can reduce greenhouse gas emissions by 75% compared to conventional systems.

India has opened up this challenge to a wider audience, launching The Global Cooling Prize – a global competition to completely re-think how we cool the spaces in which we live and work. With lots of skin in the cooling game, we're going to see many cooling disruptions come out of India. The first round had two winners – aircon giant Daikin with partner Nikken Sekkei Ltd, and team Gree Electric Appliances, Inc. of Zhuhai with partner Tsinghua University. Both winners showcased breakthrough technologies with five times less climate impact than conventional AC units. Daikin also believes that HFC-32 (R32) is the most balanced refrigerant in terms of safety, energy efficiency, economy and the environment. This plays right into its Environmental Vision 2050, which has a target of reducing greenhouse gas emissions to net-zero by 2050.

As for the cooling of large events, one will have to see how Qatar's 'outdoor air conditioning' effort works out for FIFA 2022. As the Washington Post aptly puts it, in wealthy countries like Qatar, 'climate change is merely an engineering problem, not an existential one'. But this isn't a luxury sub-Saharan Africa can afford, where heat waves and droughts are already making life almost unbearable. One can just hope that cooling will be high on the agenda of reasons to solve cheaper, more accessible energy. Else, those who live in corrugated tin houses with summer temperatures north of 45°C will only have rudimentary (yet super clever) solutions like the Bangladeshi zero-electricity Eco Cooler – made from plastic bottles.

PROBABILITY

SCIENCE

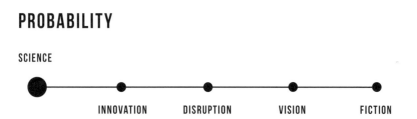

INNOVATION   DISRUPTION   VISION   FICTION

How are we doing on the Net-Zero front? The goal to bring global energy-related carbon dioxide emissions to net-zero by 2050 is the only chance we have to limit a global temperature rise of 1.5°C. Staying on track means that every possible available clean energy technology should be deployed – electric vehicles, nuclear, hydrogen, biofuels, solar, wind, the whole shebang. Worldwide clean energy investment will need to more than triple to around $4 trillion by 2030. For solar power alone, it's the equivalent of installing the world's current largest solar park every day.

If we have unlimited energy, it also means we can theoretically have unlimited fresh water. Desalination has always been thought of as incredibly energy intensive. But with the right mix of renewable and/or nuclear energy and leading technologies like multiple-effect distillation (MED), this might just be possible in 2050. MED is a proven method to produce distilled water with steam or waste heat from power production or chemical processes. This method boasts very low electrical consumption (less than 1.0 kWh/m^3) compared to other thermal processes such as multi-stage or membrane processes (reverse osmosis). It's easy to install, has very low maintenance and can operate 24/7 with minimum supervision. If you're generating round-the-clock power, does it not make sense to turn that waste heat into 24/7 water distillation too?

We will continue to innovate renewable energy. Nothing will be a sacred cow. Take wind turbines. They are super expensive and, although eerily beautiful, a bit of an eye sore. Then German start-up Kitekraft came up with flying wind turbines that require ten times less materials to develop than turbines. Consisting of a small ground station, a flexible line and a specially designed 'turnkey kite' (basically one single rigid wing), the kite then performs figure eight movements to produce electricity and prevent the line from snagging. In the long run, one of these kites will have a wingspan of up to 16 meters and generate 500 kW.

The world will take some serious convincing to leave fossil fuels behind. But with a non-negotiable climate crisis timeline and entrepreneurs stomping at the bits to get their share of the action, the future of energy is evergreen.

FLIP
THE SWITCH

While relaxing in her self-driven car, and watching the Brussels cityscape glide by, Ella Ferguson-Linux finally has a moment to reflect. What a crazy year 2051 has been. It feels like just the other day that she was a lawyer specialising in data privacy protection. Now, she's a best-selling author. Ever since her book, You are a Cash Machine (and You Don't Profit from It), hit the virtual shelves, it's attracted equal amounts of criticism from the tech companies as it has acclaim from commentators. Leading authorities, including notable guru Tim Cook-iCloud, have hailed the boldness with which Ella has held advertisers to account. The idea for the book had been seeded many years ago, when Ella was still a law student at Stanford University. Her Master's thesis had explored the subject of data privacy – back in the days when very few controls were in place to protect people from exploitation. Even then, during the 2020s, Ella could foresee a future where pressure could be brought to bear to change things for the better. With her book now translated into nineteen languages, and with the recent online announcement of its global bestseller status in the Society and Culture category, Ella is now in high demand as a guest at international conferences. Just this morning, she attended a meet-and-greet at the Royal Library of Belgium, ahead of her next round of speaking events. Now, it feels good to finally be en route home. She glances down at the content currently displayed on the car dashboard. There's a great deal offered on a weekend escape to New Zealand. She smiles. That one's been on the virtual bucket list for too long now. There's also a discount on 4D cinema tickets to see the recently released blockchain-buster, Lake Kolyma. Perfect for next Tuesday. Oh, and VIP tickets to the 85th edition of San Diego's Comic-Con. The 80th one was such fun back in 2046. What a difference it made to have only her preferred content. Just last month, a slimming pills brand had been spamming her twice a day. Now thankfully, all further attempts have been blocked until further notice.

Once home, Ella flops onto her memory-foam couch and checks the latest online news. The 5.0 Indo-European Union's GDPR framework is soon to be finalised and will be released in just a few months. This reminds Ella of her former life as a lawyer, when she was working on privacy matters for the European

Union in the 2030s. Another news piece catches her attention. The announcement of the new Chief Ethics & Data Officer (CEDO) at VeroK, Michele Roger-Linux, a fellow classmate back in Stanford. Ella has no doubt that Michele will excel at her new job. She's known for making honourable, ethical decisions.

Ella opens her VeroK official account to publicly congratulate Michele and show her support. While scrolling through VeroK, she sees that her book is being advertised online to any citizen of the world having listed 'books, lifestyle and IT' in their centres of interest. Given the amount of positive feedback in the comments, the ad is working well. Or perhaps, her book's presence on the best-seller list speaks for itself.

Ella accesses her 'VeroK – My Data' account to check her monthly payment for sharing her data with the social media giant. She has already reached the 15-crypto-euros mark, which she can use for online purchases. She's waiting to hit the 20-crypto-euros mark to treat herself to a trip to one of Singapore's trendiest, molecular restaurants.

Ella thinks back to the unfettered era of the 2020s. Times have changed for the better since then. Brands now not only need approval from users for precious, personal information. They also need to pay them for it. Progress, at last.

TOWARDS CONTROLLED AND PAID ADVERTISING

In 2051, surnames no longer look like they used to. They've been extended with a domain name, for both practical and legal reasons, depending on your technological religion – for example Ella Ferguson-Linux or Michele Roger-Linux. For a while, the 'Big Four' were Microsoft, iCloud, Google and Harmony (the former Huawei) but Linux has emerged as the counter-culture – promoting a more ethical use of data and more control on advertising on all connected devices.

So much has happened on that front since the end of the 2020s. In 2027, the giants Facebook, Google and Apple were finally taken to court on charges of anti-trust. This resonated worldwide. Connected populations around the world had known that it was just a matter of time before such a trial would take place. With connected devices having taken such a huge place in everyone's lives and given the huge quantity of data these entities could access and trade to other companies and brands, authorities had to act. Facebook, Google and Apple were also forced to give up a seat on their board of directors to Chief Ethics & Data Officers, which became the highest paid civil servants in Europe. In 2028, Facebook was dismantled and replaced by VeroK, the result of VK's takeover of the Vero social network. Since then, Baidu has remained the only brand that survived the last 40 years. And for good reason: China has withdrawn from international technology regulation entities.

That was the last straw, after the excitement around the hyperconnectivity enabled by 5G in 2023. Since then, each new connected device purchased triggered an exponential growth of unsolicited commercial invitations. As the term 'offline' faded away and as digital advertising became the norm, any pretext and any device was seen as an opportunity to reach potential consumers. Watches, glasses, dashboards, right down to the toilet mirror — it became almost impossible to escape without leaving the grid.

In 2029, the European Commission started to take the problem seriously as it was escalating to the point of calling into question the ethics of the entire advertising industry. The issue was explosive. European Commission employees in the data privacy department were receiving tens of thousands of complaints every day about the abuse of notifications or the lack of respect for privacy. It wasn't an easy fight. But technology companies were, in the end, forced to adopt a European standard for processing personal data for advertising purposes.

European law started with limiting the number of attempted solicitations by brands. Each citizen could only be targeted once in every 48 hours by a given brand. Beyond that, the penalty was clear: blocking the merchant's identifier for twelve months, unless the end user decided otherwise (unblocking from their European identification application). And soon, after a few impressive protests in the US, the law was also implemented in North America.

This outcry became an opportunity for brands, who redoubled their creativity.

Artificial intelligence very quickly managed to shuffle its data silos and find the right moments to reach consumers. The introduction of AI to read human emotions greatly contributed to more timely targeting. This caused protests in the former European southern bloc (Italy-Spain-France-Portugal) which lasted several months.

2035 was an important milestone worldwide. After eight years of negotiations, the UN passed a new international law on data management. Moving forward, brands were forced to pay citizens, wherever they were in the world, for using their personal data. The amount was, not surprisingly, low for generic data. But the highly regulated management of health data sometimes generated more impressive transactions, and often got media attention.

For example, the Mountain View biotech company 23AndMe sold data for a record-breaking 14 billion euros in 2030. Nearly 20 years of Google research associated with the DNA sequences of millions of users (Americans, Africans and Europeans) went to an insurance company in order to refine an AI capable of anticipating disease. The data, of course, had to be anonymised before processing.

In 2042, DNA data was sold on a voluntary and individual basis to private research institutes for about 10 crypto-euros (or a thousand 'old' euros). The Verok network history could sometimes fetch up to one crypto-euro per year for people engaged in paid daily status programmes. As for Amazon Wish lists, their selling price dropped to 0.001 crypto-euro after tens of millions of lists were published online by North Korean hackers. Today, official valuations have put an end to the speculation, despite a fierce reluctance from banking ins-titutions. And for good reason: their scalping algorithms were happily taking advantage of the soaring price of data, especially in Africa.

With the law forcing companies to become greener and more ethical, in line with environmental objectives, some were able to use the European Union's seal of approval in their communications. This label allowed them to commu-nicate once a semester with consumers in their electronic mailboxes, thanks to a certification in the blockchain. Few abuses have been observed in Europe. Even China quickly adopted open-source technology to create the 'Great Wall of Brands' in 2042 – the only ones allowed to communicate with its citizens. With GDPR version 5.0 scheduled for 1 January 2052, the Indo-European Union is once again pioneering data privacy by offering a strict framework for paid communications between citizens, wherever they may be on the planet, as well as companies with compliance certificates. The Americas (the merger of the United States, Canada and Central America) should adopt the same legislation in 2055, adding their various geographical bases on planets where official recognition has been ratified by the UN and where tourism is permitted to populations vaccinated against Covid-50.

DATA PRICE: THE NEXT BIG SPECULATION?

Advertising messages haven't been aimed at the masses for a long time. Even without looming privacy limitations, a shift is already taking place towards emotional connection and a bigger focus on consumer psychology. Algorithms will have to become more and more adept at seducing the narrative through emotion. A lack of access to personal data could also sound the death knell for traditional, creative advertising agencies, as they'll no longer have access to data that allows them to reach consumers with sufficient precision to be profitable. Without personalised data at their fingertips, without vanity metrics like views, impressions and clicks, marketers will have to come up with many more niche campaigns to appeal to a broad market set.

Ethical and ecological concerns will soon cease to divide Western societies. They will become widely adopted, particularly under pressure from consumers who are concerned about both the environment and the life cycle of the products they buy. Credible green claims will win customers over.

The idea of a mature Chinese operating system like HarmonyOS could boost a market where Google and Apple have held a comfortable duopoly for the past decade, and after the failures of Microsoft, Palm and BlackBerry (plus, more anecdotally, Corel). If the project attracts other manufacturers, Huawei could succeed in building Western consumer confidence in Chinese Clouds in the global battle for data.

According to a study, the global data monetisation market is expected to be worth $370.969 million by 2023. Consumers will undoubtedly understand the value of their personal data and monetise its use. Governments will then have to take measures to regulate these transactions.

This will also apply to the post-digital transformation of advertising, as it feeds on data to appeal to users. The opening of an investigation against Google by the European Commission in June 2021 is only the beginning of a campaign against the use of personal data for advertising targeting by certain technology companies, initiated in particular by more ethical players (Vivaldi, Duckduckgo, Mailfence, Tutanota).

Let's end on an anecdotal note. In England and Wales alone, almost 200 000 surnames have disappeared in the last 115 years, according to the Daily Mail. The reason? Probably since when two people marry, one takes the partner's name. Fast forward another thirty years, and we may well run out of surnames. Since technology will define and own our identities, who knows, maybe people will incorporate their chosen operating system in their last names. Tim Cook-iCloud? Ella Ferguson-Linux? Who will you be?

PROBABILITY

SCIENCE

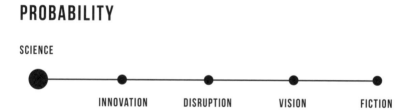

INNOVATION   DISRUPTION   VISION   FICTION

Hyperconnectivity is nothing new, even in 2021. With the surge in connected devices, imagining a world where brands can reach you anywhere and any-time with advertising, even in your toilets, doesn't seem far-fetched. It's already the case when you're scrolling down your Instagram feed. With all the possibilities that connected devices will bring, new laws will most likely have to be established to frame advertising. Otherwise, it's easy to imagine how intrusive it could become when you receive endless notifications for products or services which don't interest you. As a result, marketing will be more interesting and beneficial to everyone, and we'll return to true creativity based on data.

However, whether the Facebook empire will be dismantled by 2051 remains to be seen, as we still base a huge part of our lives on it. It might evolve into a more ethical version but vanishing completely is, for now, difficult to imagine.

DOCTOR AI WILL SEE YOU NOW

A typical morning in the year 2051. A slender woman, let's call her Mina, wakes from her slumber, pulling off the EEG sleep mask that induces lucid dreaming and measures brain activity. It syncs with her olfactory transmitter emitting just the right aroma into her bedroom to soothe her emotions and energise her for a big workday ahead. Today's scent: orange, sea-salt and wasabi, pre-programmed fragrance notes to match happy memories from her childhood. Later today, the sleep mask will provide her with a guided meditation to wind down, based on her interactions throughout the day.

As she walks to the bathroom, she rotates her smart earring to get an accurate reading of her vitals. It's the latest wearable supplied by her chosen health provider, Nadilla. The size of a small coin, it measures everything from her heart rate and temperature to her blood pressure and blood oxygen levels. It constantly monitors and shares her information with the Nadilla Care Hub – an insurance-cum-health network of connected care providers and AI-bots, powered by wearables and smart home devices.

While Mina relieves herself, a music track with subliminal motivational messages plays. The lighting is automatically adjusted to exactly the right level to coax her into waking according to the alertness level detected by her sleep mask. The daily news loads on the smart mirror above her sink. It also displays a suggested workout routine (she's been neglecting her core training), tips on improving nutrition and warnings on UV and pollen levels. Before flushing, her smart toilet analyses her urine, reporting that she needs to up her calcium intake and that she will be ovulating in three days' time.

An AI-enabled camera fitted inside the plumbing discreetly captures photos of her stool to track her digestive health. It also closely monitors other Nadilla customers in her building and neighbourhood to quickly detect disease outbreaks or other worrying trends. Mina doesn't love the idea of her most intimate details being analysed, but it's one of many necessary checks she does to enjoy unlimited free health services on the Nadilla plan.

Before brushing her teeth, she spits a small saliva sample into a Nadilla lab kit to instantly test for thousands of diseases on their database – prioritising outbreaks prevalent in her area. If an infection is noted, it automatically fires off a message to her local pharmacy to dispense the necessary medication straight to her door. In case of something more serious, a chatbot will schedule a video call with her in an hour's time to discuss additional treatments. Equipped with deep learning algorithms, the AI-empowered bot will interview Mina about possible symptoms, then refer her to a human specialist. She's had a skin cancer scare in the past, so she will be asked to scan a photo of her skin ailment into an app. This will put her in direct contact with a dermatologist if it has to be treated or removed.

Mina is going through a painful break-up, so her mental health is closely monitored by an AI speaker that engages her in conversation and checks for signs of depression. Today they are chatting about Mina's weekend plans to travel by submarine to a sub-terranean island retreat. The bot reminds her that her mandatory one-month check-in with a life counsellor is coming up the following week.

For breakfast, Mina whips up a smoothie of bio-engineered supplements – custom created for her by Nadilla – while her smart fridge suggests (and orders) her perfectly nutritionally balanced dinners for the week. Finally, when she steps outside, she straps on a breath biopsy device to track biomarkers in her exhaled breath for early detection of cancer, and if the chest infection she had a few weeks ago has settled down. At the same time, the mask will detect any hazardous airborne pathogens that might make Mina ill.

All these checks take her almost no time at all, and while she gets on with her day, data in the form of images, sounds, emotions, DNA, vitals and other indicators are shared and analysed across a massive 'health flight centre'. Welcome to the next generation of proactive healthcare.

A NEW WORLD OF ALL-INCLUSIVE SUBSCRIPTION HEALTHCARE

Our healthcare system is by nature reactive. You get sick, you go to the doctor, you get treated and get better. You don't speak again – they don't check back in with you and you don't really take responsibility until the same thing happens again. For decades, this economic model has worked for healthcare professionals because they make more money the sicker patients are.

But this isn't very user friendly, is it? It doesn't make us healthier and more conscious as a society, or incentivise us to take better care of ourselves.

What if this model can be completely turned on its head? We predict that, by 2051, we will no longer pay doctors to cure us when we're sick. Instead, all treatments will be free. We'll pay an ongoing subscription to a healthcare provider, who will offer a completely holistic mental and physical health service. We'll choose this provider very carefully – much as we take time to compare internet or mobile phone packages. Once decided, we'll commit to wearing their wearables, kitting our homes out with smart devices and being injected with their nanobots (designer medicine).

Virtual assistants and human health experts will monitor our vitals 24/7, sending us tailor-made food recommendations, sleep alerts and specialised workouts, much as we described with the Mina example. They will remotely tell their nanobots where to go for interventions and check-ups. If we get sick anyway, treatment will be covered by the health provider because they failed in their service.

While this model might sound new or even ridiculous, it's formed the basis of Chinese medicine for thousands of years. The prevention of disease and the maintenance of health the Chinese way is a total lifestyle, where you pay a doctor a retainer when you are well, and you don't have to pay when you're sick (since the doctor failed you). Through acupuncture, herbs, diet and lifestyle guidance, the doctor will nurse you back to health – and then you'll start paying again.

By encouraging patients to be more involved in their own care through constant health monitoring, it will be much easier to intervene early on – before a condition becomes acute. Imagine knowing the moment cancer cells start to grow, if gallstones are forming or your thyroid is suddenly underactive. Just think of the cost savings of fewer people needing hospitalisation, chronic medicine or other expensive interventions. Picture the reprieve for hospitals if new viruses can be instantly detected – even obliterated. The reward far outweighs the alternative of not knowing, and people will happily pay for a more effective model if it means all care will be free.

SOUNDS FAR-FETCHED? APPLE AND AMAZON THINK OTHERWISE

Teaming up with Berkshire Hathaway and JPMorgan Chase, Amazon sadly failed to get its healthcare project Haven off the ground. Not giving up, the company opened up their Amazon Care programme to other companies in 2021. The app offers a range of urgent and primary care services, including testing, vaccinations, treatment of illnesses and injuries, preventive care, prescription requests and more (aided by its acquisition of an online pharmacy). It's working on health add-ons for Alexa too, so it's conceivable that this service will gradually spread across the US and maybe even the world. Meanwhile, Apple also trialled plans for a personalised subscription-based programme with its own employees. This primary care service would integrate data from devices like the Apple Watch with clinical care. The tech giant even took over a clinic space to test the programme, but it was halted when employees questioned the integrity of data collected through the services.

As for wearables and implants, technology is advancing at breakneck speed. Engineers from Rutgers University–New Brunswick have created a smart wristband with biosensors that monitor the counts of different cells in our bloodstream through tiny pinpricks – a bit like a FitBit on steroids. Imagine sending constant blood samples to your doctor, without the hassle of a lab test. Physicians will have a 24/7 real-time dashboard of their patients' well-being and will be alerted to any changes requiring urgent attention. Sensors like Abbott's FreeStyle Libre 3 have not only transformed life for diabetics but also shown that implants are going mainstream. This tiny arm patch automatically sends minute-by-minute glucose values to a smartphone, which the patient shares with selected people.

Health trackers are quickly progressing from wrists to elsewhere. For example, the Oura ring must be the best-looking health tracker on the market right now, giving you eye-opening insights into your body and well-being from the arteries in your fingers. Another sleep tracker, the Neuroon EEG mask, is completely open-source – ready to play with whatever IoT devices you have in your home. The quest for predictive medical data mining may well take things underground, so to speak. Scientists at Stanford University published a paper on a disease-detecting smart toilet that also examines fecal matter and urine to determine the user's health. Japanese toilet maker Toto has already unveiled a concept 'wellness toilet' claiming the same capabilities.

If all things toilet-related make you blush, you may be more excited by the completely non-invasive Breath Biopsy by Owlstone Medical. It's used to measure the more than one thousand volatile organic compounds (VOCs) found in exhaled breath, as well as microscopic aerosol particles from the lungs and airways. Both VOCs and breath aerosol represent rich sources of biological information, helping to spot conditions like cancer.

Your smartphone – whatever it will look like in 2051 – will continue to be your health ally. Take Google's Derm Assist app. Powered by Google's artificial intelligence and machine-learning capabilities, it can analyse photos of your skin and look for a match in a database of 288 skin conditions. It then presents you with some possible skin conditions, with a success rate of up to 97%.

PROBABILITY

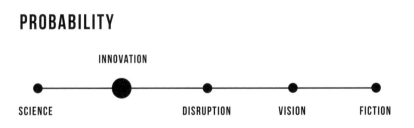

The connected healthcare model is not only designed for the super-wealthy. Technology will finally force a fragmented health sector to put the power in the patient's hands and provide quality care for all. As the World Health Organisation rightly pointed out over a decade ago, strengthening health systems is 'everybody's business'. We can't forever outsource it to professionals and not take responsibility for our own lifestyle choices. Technology will continue to evolve and we'll have a wealth of data spread across a dizzying array of devices and providers. An exciting opportunity remains for someone to bring all of this together in a sustainable model.

AROUND
THE WORLD IN 0.8 DAYS

It's Friday morning and Jolene is feeling uncharacteristically nervous. Ever since she attended that metaverse meet-up on the quayside in Darwin, Australia two weekends ago, her life has turned upside down. Her healthcare provider keeps pinging her alerts – terse reminders that her recently uneven heart rate needs to be seen to straightaway. The smart screen on her desk hasn't held back either – apparently her day-dreaming time was up 47.95% this past week. Her wardrobe keeps suggesting outfits in shades of red with some interesting hemlines. And the home fridge keeps proposing ridiculous mid-week supper ideas, in an irksome, nudge-nudge-wink-wink electronic voice that Jolene hasn't heard the appliance use. Last Tuesday it was cryogenic oysters, glacier-grown strawberries and an organic Prosecco. Every time she arrives home, the lights dim, soft music cues up and she almost trips over the cat. Things are a mess. And they have been from the minute she locked iridescence on Noah, the Australian marine biologist. On Monday, there'd been a message from him to say he'd be in New York tomorrow for an old-school, face-to-face conference, and then coming on to Florida that evening to take samples of the mangroves for a new sustainability project. The AI travel agent has sent her his itinerary. Could this really be happening?

Friday morning. Darwin, Australia. Noah tips his suitcase lid down and locks it with his fingerprint. He takes a last look at the itinerary flashed in front of him and steps out of his front door, just as the A.I.ber glides up to collect him. His A.I.ber driver doesn't lack for small talk, so the fourteen-minute journey to the speedport is quick, leaving him with ample time to catch the 7.42 am to Brisbane. Inside the speedport, a HoverCart takes him straight to platform 42. Noah settles into his seat and watches the scrubland and gum trees of the Northern Territories glide past him as the maglev train reaches its 950km per hour cruising speed. He can still recall the first train trip he'd ever taken, as a six-year old to Alice Springs – an overnighter back in '31. He smiles. Then he remembers the nostalgi-feature on his armrest that mimics the sound of a 'real' train when activated. He selects the quaint, Thomas the Tank Engine emoji. In seconds, his headset is filled with the clitter clatter of his boyhood travels, the seat gently reverberates, and Noah is sound asleep.

Three hours later, the TrolleyBot is gently prodding him awake. There's just enough time to admire the Queensland scenery and enjoy a protein-enriched smoothie before the maglev train glides into Brisbane. Once at the Brisbane speedport, Noah boards a hyperloop shuttle to the airport. He recalls his parents telling him how they used to have to be at the airport two hours before take-off. That was before biometric passport control and superfast luggage scanning. Now the check-in time is less than fifteen minutes, and when the travelator gets him to his flightcraft, he is welcomed aboard by the AttendantBot Stacey565. There's an on-board announcement that – as the flightcraft has just landed from Singapore – the hydrogen fuel levels are being checked before take-off.

As the flightcraft lifts and bends its way up and out across the Pacific, Noah is disappointed not to be sitting next to an actual window, but he's chosen this airline because the windowless craft can cut five hours off the flying time to JFK – down to four hours, thirty-seven minutes. A nine-odd hour journey, door to door. Which should get him to his conference in time for the 10 am start. And besides, all around the interior, constant updates and footage of the route are played on the window-shaped monitors. It's a nice touch for those who hanker after the Dreamliner days.

As the flightcraft enters Hawaiian airspace, Stacey565 serves up lunch, along with a small shot of FliteElixir to combat jetlag. And then, in what seems like no time at all, the flightcraft is dipping and banking towards the Manhattan skyline. How small the Empire State looks on the monitors, next to its neighbouring skyscrapers. It's almost the fiftieth anniversary of 9/11. His grandfather had been working in the North Tower that day – seconded from the Sydney office – and had only just made it out. He snaps a pic to send to Grandpa and offers up a short prayer of thanks.

Once on the ground, Noah takes the subtramway to his conference venue on the Upper East Side. The journey has been shortened, thanks to advances in battery-powered, underground tram engines. The conference is due to wrap up at 5 pm, which should be ample time for him to catch the 6.07 pm quick-hop to Palm Bay, Florida. As he takes his seat in the auditorium, he checks his antique Rolex Submariner – a gift from his father – yet again. Was it a bit too soon to be suggesting an actual meet-up with Jolene this evening? Only ten hours left to find out.

7.30 pm, Palm Bay. Jolene makes her way to the SmartHeliPort at the top of her building. There's a HeliTaxiV1 already waiting, and at her voice-command, the door opens, she climbs in, and the craft takes off for the airport. The monitor on her seat has detected from the voice command that her throat is dry, and a bottle of filtered water is lifted out of the fridge for her on a small, robotic arm. She sips it, gratefully, and watches as a small, domestic flightcraft dips, banks, and lands expertly on the recycled plastic tarmac. Looks like it could be the one from New York. She scans it with her phone camera and receives the flight number. It's a match.

A quick hop on the speedevator, and she finds herself at Arrivals. There's a familiar figure on the other side. She feels a knocking in her chest, and the bracelet monitor goes into overdrive. She looks down at her latest heart rate readings. With a smile, she de-activates the device. There will be recriminations of course, when she puts it on again after the weekend. But that's OK. She'll take her chances on this one.

PLANES, TRAINS AND HYDROMOBILES

In the early part of the 21st century, and in the first shift away from petrofuels, electricity was championed as the cleaner alternative. Electricity was mainly used for short-haul drive vehicles. However, its shortcomings prompted the development and usage of great petrol alternatives like hot air, hydrogen, helium, and biogas. Over time, all the heavy goods vehicles (HGV's) for long-haul trips were powered by hydrogen and helium.

Biogas was used well into the early 2030s. This was a mixture of various gasses, mainly methane and carbon dioxide produced from raw materials like food or agricultural waste. Its initial purpose was that of providing heating and electricity, but it was soon mostly used for the production of green hydrogen. Over the decades, due to the rise of autonomous vehicle taxi services, mobility as a service (MaaS), and shared shuttle services (including mini-shuttle buses), there was a massive drop in car ownership. By 2051, it's become a rarity for anyone to own a car. Those that do are unlocking it, using either their biometric data or the power of their minds. Biometric data was popularised with the usage of smartphones, but car manufacturers went a step further to ensure users' vehicle safety. Such vehicles are controlled via a device attached to the user's head. The device measures brain activity and translates it into car actions. The device placement procedure is non-invasive; consisting of solid electrodes that are easily attached to and detached from the skin. This gives the user plenty of options to control the car via the device, from unlocking to driving. Now, disabled people can drive just as easily as their able-bodied counterparts.

In the case of mini-shuttle buses, passengers would use the app to track shuttles' location in real-time, with each being a few minutes apart. If the shuttles are going to the same destination, e.g. airport terminals, they would eventually attach themselves to one another, and then start detaching on different terminals. If electricity reserves were running out, shuttles would simply drive themselves to the nearest wireless charging station.

The charging stations were a challenge to develop, as they had to protrude high enough not to waste electricity. The solution was to design flat stations that would slowly protrude upwards only once the vehicle was parked on top of it, and until the contact point at the bottom of the vehicle was reached. Thus, vehicles of various heights were able to use it seamlessly.

Individual car owners had their own battery charging stations. And with each car coming with two batteries as standard, battery charging, and replacement, was a simple matter. One battery could be charged at home, while the other battery could be exchanged for a full one at a battery replacement centre, with no valuable time wasted.

By the middle of the century, air travel had undergone significant changes. In order to cut airfare costs, airplane companies started producing planes without windows. Windowless planes turned out to be a much more sustainable solution. Not only were structural weaknesses eliminated, making them safer, lighter, and faster, they were also cheaper to produce and needed far less energy to run. Windows were replaced with interior screens. Users could choose between several viewing options, from daytime to nighttime to entertainment mode.

Zeppelin-like crafts, called blimps, gained in popularity due to their cost-effectiveness. Unlike a zeppelin, with its rigid metal frame, a blimp is inflated like a balloon, and its shape can therefore be more flexible. Although more expensive than hydrogen, helium was a less combustible gas, and it became widely used in blimps. Combustion engines were replaced with hydrogen batteries, which increased their safety. However, hydrogen was still used in heavy duty vehicles such as planes and road cargo vehicles.

Much like autonomous vehicle networks on the road, by 2051 we now have airborne ones, with numerous autonomous aerial vehicle (AAV) taxi services to choose from. Scheduling a craft reminiscent of the earlier helicopter is now as easy as once calling an Uber. At first, these were remotely controlled by operators in control centres; after which they became fully autonomous with no risk of human error. Human intervention was needed only in cases when the vehicle's internet network stopped working and had to be temporarily switched to another provider's 5G network, thus taking advantage of the network's flight data collection capability.

Some of these AAV companies collaborated with road autonomous electric vehicles (EVs). Since AAVs could only travel long-haul, these were used as taxis, which would move and attach passenger capsules to the nearest EV taxi without a capsule. This concept was proven to be highly efficient in busy metropolitan cities. Each building's rooftop was equipped with multiple 6-place passenger capsules. Once the trip was executed, AAVs would deliver back the capsule to the nearest empty rooftop capsule space.

Personal AAV usage was restricted to hoverbikes that were also used by delivery companies for small and lighter goods. AAVs were first popularised on the Mediterranean coast and the Middle East, where plenty of houses already had flat roofs that could be used as perfect landing and storage space.

By 2051, railways use magnetic levitation, or maglev systems. Rails are placed above the earth's surface and supported by columns. Trains move either on top of the rails or are suspended under them. Maglev technology involves the use of two sets of magnets. One set pushes the train up off the track, while the other moves the elevated train ahead, causing a lack of friction. These railways were used for passenger travel, whereas underground rail (tube) systems were mainly used for freight. Tube systems are essentially the same as maglev but sealed, which allows for minimal resistance and friction, due to the absence of air.

Ever since the 19th century, when one of the earliest designs for a spacecraft was conceptualised by Konstantin Tsiolkovsky, humans dreamed of space travel. Twentieth century scientists used rockets to travel into space. The cost of these spatial expeditions – combined with the cost of rocket pro-duction – meant there was a need for a more sustainable solution. However, once we had acquired the capability to visit Mars in the early part of the 21st century, space travel became an exciting new space for innovation. By 2051, the Chinese may well have completed the construction of a full-blown, colos-sal space elevator, extending 35.000 kilometers from the Earth's surface and reaching geosynchronous orbit. It's predicted that the USA and Russia will be building their versions. Two main reasons abound for wanting to reach the top of this elevator. One is space tourism, while the other is space exploration. However, in 2051, space exploration is reserved exclusively for members of the (CNSA) China National Space Administration.

STAIRWAY TO HEAVEN? OR AN ELEVATOR?

There's ample proof that the world is moving towards electricity as the more sustainable solution. However, its shortcomings, like storage, and its inability to replace fuel for heavy or long-haul tasks, are becoming more apparent by the day. A great example of a failed attempt to support electricity production was that of the solar panel roads that used one-third of the electricity they generated, just to power their built-in LED lights to increase nighttime visibility. Their creators didn't consider leaves, snow, or heavy traffic covering the roads either.

This is where green hydrogen comes into the spotlight. Compared to 31.9 terawatt-hours of electricity generated by biogas in a year, biogas could generate almost double the amount of hydrogen, totalling 58 terawatt-hours via steam reforming. Steam reforming is a method of producing hydrogen by the reaction of hydrocarbons with water, whilst using biogas as the feedstock. Biometric data is already a part of car purchase packages. Already, Hyundai Santa Fe owners can unlock their vehicles with fingerprints instead of keys. Multiple drivers can register their fingerprints, and depending on the fingerprint in use, the car will automatically adjust seat positions as well as the angle of the rear-view mirrors. Being able to do so only using the mind is not too far-fetched either. Mercedes-Benz is trying to make it a reality. In 2020, they unveiled the AVTR Vision car that recognises drivers by their breathing and heartbeat. The owner of the car can wear a device with electrodes at the back of the neck that reads brainwaves and translates them into car actions. The car is not yet for sale, but the prototype's test rides look promising.

Mobility as a Service is undoubtedly on the rise, and so is the development of autonomous vehicles. Plenty of autonomous driving technology development companies are trying to be the first ones to put fully autonomous vehicles on the road. Such vehicles are Waymo, Zoox, Zoe, Apollo Cruise and Motional, with big companies providing funding, such as Google, Amazon and Baidu. These autonomous EVs – also called robotaxis – are still in their testing phase. The S3 or shared shuttle service is a pilot project by Keolis, testing autonomous electric vehicles in the city of Gothenburg on behalf of local public transport authority Västtrafik. The shuttles, made by Navya, have a maximum speed of 20km/h and can carry up to eight passengers. Charging takes five hours, and the vehicles can run for up to seven hours on one charge, depending on weather conditions.

Wireless charging is something BMW already offers. The charging system consists of a charging pad or station, with the primary coil to be installed in a garage or outdoors. The secondary coil is under the vehicle. An alternating magnetic field is generated between the two coils, through which electricity is transmitted without cables or contacts at a charge rate of up to 3.2 kW, allowing for the battery to be completely charged after three and a half hours.

Battery swap stations are something already being developed. As of today, car producer NIO has built 301 NIO Power Swap stations, 204 Power Charger stations, and 382 destination charging stations in China and completed more than 2.9 million swaps and 600 000 uses of One-Click-for-Power services. The reason our vision predicts two batteries in the cars is that you must swap your initially perfect battery that comes with the car, with the one that perhaps doesn't have a 100% battery health anymore and will hence drain faster. Keeping the ownership of one battery could make us more likely to hop on the battery swapping trend.

Though we're so used to flying planes with windows, it might become a thing of the past. In 2018, Emirates Airline unveiled a first-class suite in one of their Boeing 777 aircraft that features virtual windows that project images from outside of the aircraft. Emirates President Sir Tim Clark regards the endeavour as the first step to windowless planes.

Zeppelin-like aircrafts are making a comeback. British company HAV (Hybrid Air Vehicles), announced plans to enable short-haul flights for city-hopping in their Airlander 10 blimp. HAV calculated less than a tenth of the carbon footprint created for the same journey conducted with a blimp, than with a conventional jet plane. Their blimp uses helium as a lifting gas. Other companies heading towards blimp production include US company Lockheed Martin, French Flying Whales, and Israeli Atlas LTA. Even Google's co-founder Sergey Brin is investing $150m into the creation of a 200-meter blimp, expected to be the largest of its kind, with some sources claiming it's to be used as a luxurious 'air yacht', and others saying it's to be used to deliver necessities to remote locations on humanitarian missions.

Toyota, Uber, Hyundai, Airbus, and Boeing are some of the companies working on the development of flying taxis. A select few are already working on autonomous aerial vehicles or AAVs. One such example is Ehang's Ehang 184. This 1.77-meter craft resembles a large drone, can reach 130km/h and is made to be eco-friendly, secure, autonomous and suitable for short and medium-haul transport.

Combining EV taxis and AAV's is also an existing concept. Audi, Airbus and Italdesign's Pop.Up Next includes a passenger capsule that is delivered by a drone to the nearest taxi available. The concept is currently being showcased and tested using 1:4 scale models.

We could also soon see bicycles moving from the roads to the air. UK's Malloy Aeronautics and Russian Hoversurf have already developed hoverbikes. Hoversurf has moved fast from prototyping to commercial use by revealing a deal to sell the bike to the Dubai Police and later passing US federal Aviation Administration ultralight vehicle classification requirements.

While only six commercial maglev trains are operating at the moment, it'll be interesting to see whether there will be more of them due to their energy efficiency, quietness, reliability, high speed, and low maintenance. There are three in China, two in South Korea, and one in Japan. A type of maglev evolution system is Virgin's Hyperloop, which is different to the traditional maglev system in that it's a sealed tube system, removing additional air resistance. Virgin Hyperloop is currently being tested in the desert outside of Las Vegas.

And finally, a space elevator is being considered at the moment as well. NASA claims the concept of the elevator is sound, and communities of researchers around the world have shown optimism towards building one. China wants to build one as soon as 2045. A research team at Tsinghua University in China has patented the technology and published part of their research in 2018, where they explained how the fibre they created could be used to build the space elevator. According to the team, one cubic centimetre of the fibre, which is composed of carbon nanotube, would sustain the weight of over 800 tonnes.

PROBABILITY

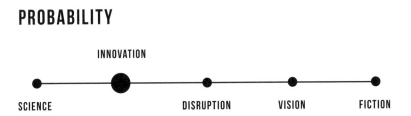

INNOVATION

SCIENCE   DISRUPTION   VISION   FICTION

One thing is sure, innovative transportation solutions can be rapidly developed, but implementation is incredibly slow in contrast. Transportation is heavily monitored by governments and influenced by the governing parties representing their interests and those of the population. And on top of the challenge of creating vehicles and systems safe for the users, there's the need for them to be safe for the environment.

The majority of the technology mentioned is available but not mainstream yet. Or it's prototyped but not yet properly funded for mainstream production. Maglev is a great example of a system introduced in the 1940s, which has not received the attention it deserves. It is possible that Virgin's or SpaceX's Hyperloop will popularise the concept. However, most governments consider it more cost-effective to repair existing railways than to replace them with maglev. Which is true, but only for the short-term, as the costs of traditional railways for users, governments, and the environment quickly add up.

Innovations on the road are more likely to come sooner, like autonomous personal vehicles and taxis. But there are some reservations regarding aerial vehicles. Firstly, we'd have to sacrifice either usage of drones, personal aerial vehicles, or taxis as there's just not enough space in the air for all of them. And since there are no rules in place on how to regulate air traffic roads above cities, it would be hard to design aerial vehicles to fly autonomously. There's also the lack of landing areas. There are just not enough flat, sizable places where crafts can land. The aircraft, in general, are also very loud and would lead to noise pollution. Powering them on electricity is not an option either as they would need too much of it to fly. On the other hand, we'll be seeing a lot more aerial vehicles, such as are the zeppelin-like blimps.

A superstar element not gaining the attention it deserves due to electricity taking all the spotlight, is hydrogen. Due to misconceptions and inadequate education on how it can be sustainably produced and stored, it is routinely overlooked. Data on hydrogen is widely available. But perhaps, with so much invested in electricity, there may be too much at stake to find a replacement. And, although there's scepticism surrounding our ability to take an elevator to space, we'll definitely be able to do so, but probably not as soon as 2050.

THE ANGEL
IS AN ALGORITHM

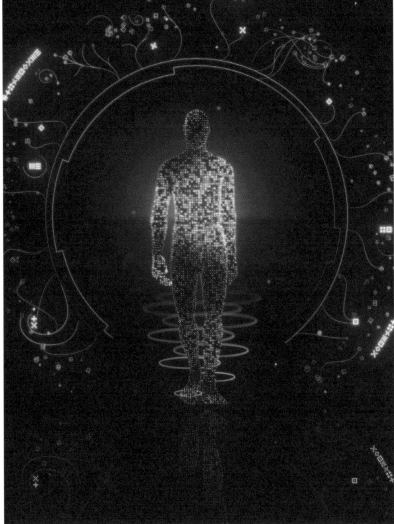

It's a Tuesday, the year 2051. Gabe orders an iced caramel macchiato at his favourite coffee place and reaches his office at 8 am. Aged 35, Gabe is an extreme sports fan and proud father of a little girl, Lucy. He heads up cyber-security at Nafios, the smart pill giant based in New York. And when we say giant, we're talking more than 80 plants with completely automated production lines. Gabe pilots this whole realm and ensures the safe production of millions of little gums that protect humanity from diseases.

He fires up his holographic computer and gets the reports HAL, his AI assistant, made on all the events that took place on the production chains of each of the 80 plants. Usually, these reports are quite fast to read thanks to the company's ultimate cybersecurity system. But today is different. Thousands of kilometres away from New York, in Buenos Aires, one of the plants avoided a catastrophe thanks to AI agents.

It's 3.37 am in Buenos Aires. Felipe, the local cybersecurity director, wakes up with a jolt. A terrible alarm is coming from his phone. It's red alert – something's seriously wrong at the plant. The machine dosing paracetamol in the Nafios smart pills seems completely out of control. The AI agent who did the seventh test of the night on a smart pill sample sounded the alarm. It calculated 4000mg of paracetamol instead of the supposed 1000mg. A very dangerous dose that could lead to liver failure, haemorrhage or cerebral oedema.

When Felipe reaches the lobby 15 minutes later, his staff is on the warpath. All the smart pill batches that exited the production lines in the last 30 minutes have to be destroyed. While this is happening, the cybersecurity AI is already doing an analysis of what happened – thanks to his Threat-Intelligence-as-a-Service solution, running on Ethereum. The result comes out in just a few seconds and the threat is blocked in less than a minute.

As Gabe goes through the report, he can't help but be very impressed by the modus operandi of the hacker. The firmware was modified remotely by the attacker, who impersonated an operator thanks to his quantum computer. He is very strong: he could even crack the operator's connection encrypted in sha256. But HAL outsmarted the hacker since it can access the whole threat intelligence system of the Ethereum blockchain it is connected to. The attack path was uncovered in record time, the attack blocked, remediation solutions set up fast, and the whole production chain restarted in just one hour.

HAL highlights the vulnerability points with best solutions to implement with a priority tag assigned to each issue. Priority: changing the encryption algorithm that has been compromised. And here again, HAL directly chooses the best alternative and prepares the re-setting with an AES encryption for all the production chain devices in Buenos Aires. This includes all the connected machinery, the hardware of the operators and management staff, the badges to enter the lobby, to name but a few. HAL will ask for Gabe's approval to launch the deciphering and re-ciphering process. To confirm the approval, Gabe's face is scanned by his computer to ensure maximum security.

Right after, Gabe starts a PKI-encrypted video call with Felipe to let him know about the changes that will happen and, above all, to inform him that he'll have to approve the local AI encryption action – again, by face scanning. Once both approvals are gathered, HAL shares its knowledge with the local AI assistant who initiates the ciphering process.

To close the case, Gabe makes sure that all the reports and findings of the AIs are stored following the InterPlanetary File System (IPFS) protocol. While doing this, he takes the opportunity to check all the previous PGP-encrypted files about incidents that happened since 2040 (when he joined the company). He also asks his AI assistant to run a matching test to check for a hypothetic correlation.

When HAL's done working, Gabe gets a PGP-encrypted report only he can access. The attack is looking just like one that happened two years ago – in November 2049. This time, the attacker messed with the packaging system at the Berlin plant. Consequences wouldn't have been as dramatic as changing the dose of paracetamol, but it had still cost the company a lot of money to fix. Thanks to threat intelligence, HAL could cross-check the two attacks and fine-tune his recommendations for future cybersecurity improvements – especially in firmware's protection. When Gabe gives his approval for these recommen-dations, the AI assistant will set everything up. It will run a test by trying to hack the system to check if all the new setups are protective enough. Case closed.

It's 5 pm in New York. Gabe has to leave work early so he puts the safety of the company in HAL's capable hands. With a long career in cybersecurity, Gabe has spent time with many companies. Over the years, his role has changed a lot as technology – especially with the help of AIs – has become more and more efficient in countering and blocking attacks. His role remains key for Nafios but thanks to his AI assistant, halting sophisticated attacks from smart hackers has never been easier or the cybersecurity system this strong. Who knows what'll come next? Hopefully hard times for cyber attackers...

AN IMPREGNABLE FORTRESS

Cybersecurity has been part of our lives for several years now. 30 years ago, companies were already well aware of the danger posed by cyberattacks. They implemented protocols which were considered strong at the time. But in our private lives, cybersecurity wasn't a big thing. Of course, users of connected devices were always happy to know that their data was protected and that some of their conversations were encrypted. Besides that, no-one paid much attention to it.

But as online crime increased, cybersecurity became more of a concern. It started first with new device-specific privacy services encrypting users' browsing. On social media and on all apps where information was shared – e.g. messaging platforms and banking related apps – PGP, AES or PKI encryption has quicky been generalised. By 2030, users' private lives were already far safer than they used to be when social media boomed. Viewing pictures, or any other personal information from a relative, became possible only when you had the right private key.

With double encryption protocols becoming the norm – combined with IPFS that ensures security and inalterability of documents – even the most reluctant people started to sign, store and manage their most important files online. Everything from housing to banking contracts. Thanks to IPFS, all sensitive documents are protected from cancellation and from being compromised. That has simplified a lot of administrative procedures as people felt more secure online than ever before. They no longer felt the need to meet their advisors face to face. And they were right to do so: cyberattacks targeting users' personal data have become more and more complicated. By 2030, attacks were reduced by 54% compared to 2020.

At the beginning of the 2030s, companies stared to widely use AI for cybersecurity purposes. They've proved themselves incredibly efficient at automatically detecting anomalies and common vulnerability exposures to prevent cyberattacks. At first, AIs were only used preventively. From 2033 onwards, the first AIs were able to track down cybercriminals and block them in a defensive way. It wasn't as effective as it is now, but already a big step forward. The event that really fast-tracked cybersecurity improvements was AGA-gate, which happened to AWS, Google Cloud and Azure. On Black Friday, November 2036, the biggest cloud service provider underwent the worse attack of all time. A huge DDoS attack brought the giant to its knees for several days. The outcome was catastrophic: millions of dollars lost, and more than five billion users impacted. It became obvious that such a centralised system, even though it was energy efficient, became more vulnerable as technology improved. Quantum computers are now accessible to more people with bad intentions.

That's when experts started to work with blockchain as a new hosting opportunity. Over time, blockchain became more energy efficient. From encouraging cooperation rather than competition between miners, to using Proof-of-Stake mechanisms rather than Proof-of-Work ones, blockchain became less and less energy-consuming. So much so, that Proof-of-Work is now a remnant of the past. Blockchain's energy consumption wasn't hindering companies anymore. In 2027, the first companies took the plunge and decentralised all their data, initially cloud-hosted on a blockchain like Ethereum. Year after year, in addition to securing their data and making them almost impossible to lose or alter, blockchains were containing more and more information. They became a huge asset for cybersecurity. In 2051, when you have a cybersecurity issue, thanks to threat intelligence, your AIs and/or operators can find the attack path faster than ever.

It's gone even further. In 2041, the biggest blockchain providers decided to offer a common new service to companies: Threat-Intelligence-as-a-Service (TIaaS). While blockchain providers remain completely independent, this common service allows chiefs of cybersecurity to access the widest source of knowledge on cyberattacks ever created. It's a powerful tool that is helping companies to resolve tricky situations in record time. By 2051, this system is adopted by 93% of companies in the world, giving hackers a very hard time indeed.

In 30 years, the cybersecurity world will be completely transformed. It will be compliant with the three cybersecurity principles – confidentiality, integrity and availability – making it far easier than ever before.

TECHNOLOGY RESCUES CYBERSECURITY TEAMS

In 2021, cybercrime is dramatically increasing – something that is to be expected, considering how much time we spend online each day and all the things we can do from connected devices. Add to that the fact that IoT devices will increase to 43 million by 2023. Since the Covid-19 pandemic started, cybercrime increased 600%, costing hundreds of millions of euros. Yet, cybersecurity isn't taken lightly. Companies are all aware of how important it is to deploy robust cybersecurity protocols to protect their users and their business. Still, the concern is quite new and hackers are becoming more ingenious in their attack framework. There's a lot of room for improvement: more than 60% of cybersecurity experts agree that their cybersecurity team is short-staffed.

AIs are considered one of the most robust cybersecurity solutions to tackle the increase in connected devices and the need for better cybersecurity. As they are capable of analysing a huge volume of data, they can detect breaches and anomalies in a fairly short amount of time. They'd be able to identify any potential anomaly in real-time and give an optimal answer fast. Companies like Twenty20 Solutions are already integrating machine learning combined with real-time video monitoring to detect these anomalies as they happen. And in general, the Supervisory Control and Data Acquisition (SCADA) market share is increasing, showing that companies are moving towards a new style of solution to support their cybersecurity teams.

Encryption will also be a big cornerstone of cybersecurity. As quantum computers become more accessible and fall into the wrong hands, single password security will no longer be enough. With small quantum computers already in existence, companies like IBM are already planning to improve them in the coming decade to reach million-qubits systems. With such powerful machines, the most sophisticated passwords could be deciphered in just a few minutes (while it would take several lifetimes for the best regular computers today). That's where PGP, PKI or AES encryption systems come in. They are already used in apps we all know. For example, Whatsapp's Signal protocol ensures end-to-end encryption of our conversations.

This might go even further in the future with end-to-end encryption applied to social media. According to Ero Balsa, Filipe Beato and Seda Gürses from KU Leuven, Belgium and NYU, USA, social media could easily apply end-to-end encryption. And as it will probably become highly desirable for users, they might have no other choice in the future to keep their status of trusted authority. Coupled with IPFS, end-to-end encryption with a double key system will make official documents safer online than ever. Users will have full control over just how confidential and inalterable they want them to be. As IPFS protocol is already used in 2021, it will no doubt become more commonplace – plus combined with other technologies to ensure the ultimate security.

Today one of the big weaknesses of the Internet – and all connected devices – is the reliance on a very centralised system hosted on servers and on the cloud. Even the most minor cybersecurity vulnerability could affect the whole world and trigger the loss of huge amounts of data. The future will lead us to decentralise our online life.

A tangible track is blockchain. At the moment, blockchain is consuming too much energy compared to servers, and the biggest players fear that their energy bill will skyrocket by making this shift. But blockchain also has undeniable assets. First of all, it's quite new. By 2051, as solutions emerge, we can expect it to be less energy consuming. Proof-of-Stake might take over Proof-of-Work. Authority figures such as Elon Musk and Bill Gates are already describing Proof-of-Stake coins as 'green coin'.

But blockchain providers' big assets are tied up in the knowledge they can offer as threat intelligence. When a cybersecurity problem like the Nafios example is detected, companies can solve it at lightspeed using almost infinite knowledge of blockchain coupled with AI. However, in 2021 this solution is not yet possible as blockchain might actually cause more problems than solutions. After all, it has its own cybersecurity threats to deal with like illegal mining.

However, by 2051, bringing together all the tools we currently have – encryption, a decentralised file system, AI and blockchain – we can imagine that cybersecurity will be taken to a whole new level with almost unbeatable security systems. It then all depends to what lengths of ingenuity hackers will go to adapt to this new paradigm.

PROBABILITY

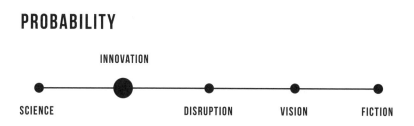

It's hard to predict the future but the Nafios cybersecurity scenario is far from impossible. As a concept, we already have the 'sketch' version of all the tools that Gabe uses against the hacker. But whether we speak of AI, blockchain or double encryption, all these systems are quite new, answering the needs triggered by an increasingly online world. Therefore, we can expect huge improvements by 2051 – things we can't even begin to imagine. Blockchain still sits with a few question marks over its head. Will it be able to reduce its energy consumption enough so that big players will consider using it instead of cloud hosting? It might be the case as there are several projects on the go that are trying to evolve towards a 'greener blockchain'. It's also important for blockchain to remain decentralised and avoid a monopoly situation at all costs.

Another question mark concerns AI and to what extent it would be able to handle a whole massive breach like the one painted in the Nafios example. Not only identifying a threat but also handling the after-effects – i.e. to reinforce security after a breach. We can assume that, combined with threat intelligence, this would be possible. But time will tell how autonomously this would happen.

Cybersecurity still has room for improvement and as the online world is an ever-evolving universe, it is surely going to surprise us in the future. Will it properly comply with the Confidentiality-Integrity-Availability (CIA) triad? Let's meet in 2051 and see.

DIGITAL
BUTLER

Cillian has been working hard to launch his own business. After two years of relentless work, he has granted himself two weeks of well-deserved vacation at his second home in Tuscany. During the two first weeks of May 2051, he had swapped the fickle Irish weather for the Italian sun. But after living la dolce vita it's time to come back to reality. As he waits for his plane to the Emerald Isle, he checks the weather, hoping it won't be that bad. And oh my, what a miracle! The famous Irish rain seems to be making way for very nice sunshine in the coming days.

One hour later, he sets foot on his native land in Galway's brand-new airport. He takes the self-driven taxi bus to Limerick and enjoys the sunny landscape. The trip isn't long – only 45 minutes – thanks to the taxi bus high-speed lane. Cillian disembarks from the bus right in front of his flat and already smells the wet mud coming from the outdoor vegetal walls. It must have rained a lot the past few days! He walks on the bright, white slabs paving the building's fore-court, right toward the entry door that automatically opens when the face-scan nano-camera confirms his identity. The hall lights up brightly when he enters. The elevator recognises his face and takes him to the top floor of the building. 11th floor, Cillian's apartment door opens automatically, recognising its owner. Inside the large living room, Kero, Cillian's dog waits patiently for him. The sweet voice of Eirebot is welcoming him back. Cillian likes to think of Eirebot as his virtual butler. He baptised it after his love for his native country. Eirebot can control more or less everything in its little 100 square metre realm and is connected to a control panel from where Cillian can check multiple data such as his latest messages, news from the flat, temperature and much more.

The apartment is as clean as a whistle despite the two-week absence of its owner and there is a nice sandalwood scent throughout – Cillian's favourite. Just before his departure, Cillian had programmed Eirebot to launch the BACKHOME programme which triggered, among other things, the vacuum-bot, and the essential oil dispenser, one hour before his return. The bot also took good care of Kero. The dog's general mood seems really good, according to the control panel display. The bot was programmed to deliver exactly 90 grams of dog-

food, three times a day at fixed times. It has also been letting Cillian's sister Annie enter the apartment every evening to take Kero to the park. Speaking of Annie, it's her birthday in a week! Cillian is lucky the information popped up on his control panel. Just a few more days to find a nice gift.

As the owner of the top floor apartment, Cillian has access to a nice rooftop terrace. He climbs the stairs to enjoy the last rays of sun. The terrace floor is made of white concrete with artificial crystal particles. With the sunset, it gives a mesmerisingly shiny spectacle that he's not used to seeing in the Irish weather. An ideal ambiance for a glass of Chardonnay. But then his phone rings. It's Greg, his best friend and co-founder of the company. Greg would like to meet with him tomorrow at 10 am to visit the R&D lab where they are developing their new, lab-grown vegan meat substitutes.

Night falls and so does the temperature. It's still Ireland after all! Cillian heads back inside. The whole apartment automatically switches to EVENING mode. The lights in the living room have become softer and the stove hood lights are now shining brightly. Time to prepare something for dinner. There's almost nothing left in the freezer, but his smart fridge tab indicates to him what he can make from what's left inside. As the tab is connected to the control hub, it also suggests a grocery list, based on what Cillian usually likes to buy, which he can send directly to his favourite supermarket for delivery. Tonight will be chili sin carne. As the cooking smoke comes out of the pan, the kitchen hood automatically extracts at exactly the right strength. The connected pan lights up green when Cillian's dish is ready. Time to eat!

As soon as Cillian sits down at the table, Eirebot starts the 4D TV. But the young businessman isn't paying attention and puts his quantum laptop on. Finally, after two weeks without any proper online protection, he's happy to be back to the online security his home provides. Speaking of security, he learns via the notifications from the AI syndic on his control panel that someone, clearly with no good intentions, had recently tried to break into the flat. Luckily, the flat was prepared for this, thanks to its own bot, which automatically called the police. How silly these thieves can be. It has become impossible to break into homes with all the technology deployed to protect it.

After a few hours on the computer, Cillian heads to the bedroom. All the lights from the other rooms are progressively switched off while the bedroom is bathed in a low wavelength light, boosting Cillian's melatonin production.

The temperature is ideal for good, restorative sleep, not too cold and not too hot. Eirebot has automatically set an alarm clock according to the time he will need to get ready, taking traffic predictions into account, to make sure he's on time for tomorrow's meeting with Greg. Cillian jumps into bed. The light slowly decreases as Eirebot plays four to seven Hertz soundwave music for relaxation. Everything is set to have him fresh and ready for tomorrow.

YOUR HOUSE AT YOUR SERVICE

In 2051, a building – especially for residential purposes – isn't what it was thirty years ago. The context is different, the climate has changed – a new paradigm now exists. With rising sea levels, flooding has become more and more recurrent, especially for maritime states, and temperatures have become even more out of whack. The need to investigate new ways of building accommodation that can withstand these new, and extreme conditions, has become more urgent than ever before.

To address the extreme temperatures, steps were taken towards cooling down cities. Up to 84% of cities with more than 100 000 inhabitants – and 100% of the biggest megacities – made it compulsory to use the most reflective possible materials and coatings on houses and buildings, thereby reducing the effects of the sun's rays. Manufacturers and DIY shops started to offer new solutions in this matter, including ultra-reflective paint, and concrete embedded with artificial crystal. Over a period of ten years, this solution allowed citizens of the world to reduce their cooling bill by 20%.

After multiple, devastating floods, many coastal and river cities worked towards a solution to stem the rising tide. In this regard, plant walls proved to be really efficient in absorbing excess water during heavy rains and controlling floods. So much so, that they were imposed on at least two walls of each new building, whether it be houses, flats or businesses – built after 2043. And while the floods became more and more persistent until 2040, these disasters started to recede as more and more buildings installed plant walls. In addition, these walls allowed considerable energy savings by acting as natural barriers in keeping the cold inside during summer and the heat inside during winter. The gardening of living walls became easier thanks to automatic watering using rainwater tank supplies, as almost every building has had one since 2030.

By 2051, these interventions – reflective paint or concrete and plant walls – have become a *sine-qua-non* condition to help us better control climate change. But the bigger revolution takes place at the heart of everyone's homes.

WATCH WHAT YOU SAY, YOUR HOUSE IS ALIVE!

2021 already saw the advent of Google Home, Amazon Alexa, Apple HomeKit and other hubs to help manage your house or apartment. But these were quite intimidatingly complex if you were a newbie. For example, you had to make sure that every device in the house was compliant with your home assistant, and there could be a danger of being tricked into buying brand new lightbulbs with sensors, that weren't compatible with the system you had for the house. These connected devices were either working on Bluetooth or Wi-Fi. In the case of the former, as soon as you lost signal you might encounter working problems with your devices. And while Wi-Fi was a bit more reliable, it required more energy for optimal working.

That's when Matter (standard) and Thread wireless mesh protocol arrived as a saviour at the beginning of the 2020s. And by 2051, it's impossible to live without it. Since 2031, every home has had its own Matter ID. When buying a new place to live, or when building your new house, integrating a complete smart home system has become mandatory. Each house ID is associated with a specific layout on the Matter Hub and all the parameters are already set as the default. It's then up to the owner to parameter everything and to use shortcuts for specific actions.

Indeed, like green walls and high albedo materials, smart homes have become more and more efficient as Matter and its Thread communication system have improved. Automatically switching the heaters off and sensors on lights was already a thing in the 2020s, but now your smart home is also your personal energy assistant. Whether it's lowering the heater when it doesn't sense your presence in the room, or automatically adapting the temperature given the amount of sunlight, everything is designed to minimise your energy bill and footprint.

Your house is also becoming a presence on its own. If you programme it that way, it will be able to welcome you with a warm fire in the stove – or just with the heaters on – and with your preferred lighting. It can even play your favourite playlist or station as you open the door. Your home takes care of you by setting the right ambiance at the right time, relaxing in the evening for a good night of sleep, energising in the morning with some positive music and a concentrated mango scent. It even goes further. Some modest income families even programme their house to automatically go to ECONOMY mode – with very little heating and low lightning – when their bank accounts are below a certain amount.

The smart homes of 2051 are also playing a huge role in security, both online and offline. Each house now provides a very strong firewall to any connected device in the house, including computers, smart TVs, cameras and phones but also connected light bulbs, fridge, oven and so on. Each house has its own AI, included in Matter's hub – capable of detecting any breach or any attempted intrusion. And it's the same in the offline world. As almost every dwelling is fitted with face-detecting cameras, the police are instantly notified when someone is trying to break in. The house is also capable of detecting all risks pertaining to electricity, gas, or fire. These security features have even become compulsory in many cities in the world.

But the Matter hub offering has also extended to include offices, government buildings, museums, warehouses and so on. It's become indispensable for practically every building on Earth. From 2045, it was the responsibility of each company and government to equip their buildings with smart home devices. Financially and environmentally speaking, it was in the interest of all to drastically improve energy management in offices.

As demand for Matter hubs was increasing for offices, the Connectivity Standard Alliance launched its brand-new hub solution for companies. The BMatter solution was tailor-made for each company. Printers on 24/7, heaters next to open windows, lights on in empty rooms, all these are now just bad memories from the end of the 2020s. Just like smart homes, smart offices are managing everything seamlessly. Companies even have energy reduction targets to respect every year and must publicly share their annual energy consumption reports. Poor performers are penalised with a fine proportional to their energy waste. 2050 was the first year where no fine was recorded around the world, and companies managed to reduce their energy bills by 43%.

BMatter, of course, also provides premium security tools. So much so that data leaks have become very rare. BMatter can protect each employee's devices and thwart hacking attempts efficiently. The Thread radio standard allows for a very rapid response from any connected device registered in the hub – even over long distances – thanks to Mesh networking.

A NEW BEGINNING FOR SMART HOMES

We are in 2021 and heaters switching off when they detect an open window aren't anything new. There are even apps that allow you to control them remotely and programme them as you wish. Apple or Google Home applications offer the same thing, but on a larger scale. It is already possible to programme your whole home to have specific lighting or heating parameters according to the time of the day or season. But the fact is that nowadays, the smart home ecosystem and devices gravitating around it, are complex. And this complexity prevents many people from adopting these new technologies.

Already there are different ecosystems – Apple, Google, Amazon – that aren't compatible. You have some objects connected using Bluetooth, Zigbee, Zwave, and others with Wi-Fi. And on top of this, you have many independent brands that sell items compatible with only part of the ecosystem. A new standard is currently being established with Matter. Yes, Matter already exists! With its hundreds of partnerships, Matter is starting from the best, open-source smart home model, Apple HomeKit. It uses the Thread radio standard and Mesh networking for a better and wider communication between the connected objects.

The Matter standard is at an early stage, but it will allow a more open access to smart home appliances regardless of your chosen ecosystem and brand. And it will go even further. With this global standard set, smart homes will be able to operate as a network and to 'communicate' amongst themselves. Let's imagine that a house at one end of a city detects a change in the weather and that rain is going to fall. Thanks to a waterfall reaction, the house on the other end of the city will be able to get ready for the bad weather and close the windows and the sunscreen without involving its owner.

We can also expect homes to become increasingly smart and instead of executing a simple order, they will be able to act completely independently, thanks to the data collected from the outside world. There's sun hitting the windows, let's lower the heater and switch the lights off. So the house can detect that its owner is on his way back home thanks to signal from his phone? Let's heat the room, switch the lights on and start his favourite music. Everything will need a parameterisation but after that the home will act just like a living entity at the service of his owner: a real digital butler.

This new standard will also pave the way for bigger entities such as businesses, government offices and public buildings such as train stations or airports to use these smart connected devices. Matter will surely help them manage their energy consumption better in a very wide area, thanks to the Thread radio standard. We can also imagine that a tailor-made offer will emerge for these new buildings and organisations.

And regarding the materials used to build the houses, the research work of Haider Taha, David Sailor and Hashem Akbari from the Energy and Environment Division of the Lawrence Berkeley Laboratory of the University of California has already proved that high albedo – highly reflecting – materials can be used for reducing energy use. As for green walls, researchers Gabriel Pérez, Julia Coma, Ingrid Martorell and Luisa F. Cabeza have showed that Vertical Greenery Systems – plant walls – have proved their efficiency in reducing energy consumption especially during cooling periods.

Combining smart materials and smart home technology will undoubtedly help humanity to manage their energy consumption in a more efficient way, whilst also combatting the climate crisis.

PROBABILITY

INNOVATION

SCIENCE   DISRUPTION   VISION   FICTION

Smart homes as we know them will surely improve with many new features and by being more accessible to all. Privateers and companies alike have a mutual interest in turning to this technology for energy management and increased security. Online security will probably be a cornerstone of these developments, and it's not far-fetched to think that houses will also be capable of providing online security. It will be a huge necessity given all the connected devices these homes 2.0 will gather.

Having a living house communicating with others will probably happen but as smart home ecosystems like HomeKit and Google Home are only in their infancies, we might not see it before a few decades.

BEAM ME UP, SCOTTY!

It's early Tuesday morning but Lauv Vandenberg is already on the warpath. He's sitting in his office at home, in the heart of Lyon, getting ready for his 2051 New Year's speech to Lauv Ltd executives. The pastel green wall behind him brightens up as the sun rises. It's 7 am, and the meeting will start in just a few seconds. Lauv puts on his connected 360 glasses, activates his ear implants and with voice command, turns on cameras located all around the room. The gong rings, announcing the beginning of the speech. Lauv and his ten executives are virtually sitting around a table. The visual rendering is completely adapting to the time and location of each participant. Lauv can actually see the sunrise reflected on the 360 glasses of his CTO, Alaric Miller, even though Alaric is currently attending the meeting from Los Angeles where it's 10 pm local time. Everyone around the virtual table takes a look at the day's agenda displayed in the centre.

Even though he stopped working two years ago, at the age of 73, Lauv always likes to give an inspiring and encouraging talk to his executives to start the year. And after his 18-minute presentation, he encourages everyone to share their thoughts and vision for the coming year. After another 30 minutes, a hologram appears and ends the meeting with a 12-minute group meditation.

Afterwards, Lauv gets ready for his sports routine. He walks to the white room next to his office, puts on his connected sports outfit and starts the session. His 360 glasses plunge him into a mixed reality world where his coach is waiting. Lauv's vitals data is very good and the AI in charge of his sports programme spices things up with 30 minutes of HIIT training. His connected sports clothes collect every bit of data on his movement, performance and vitals, from which a complete report is filed at the end of each month.

What a sweat! Lauv takes a quick shower. On his return, the fitness room makes way for exploration. He finds himself at the Royal Palace in Copenhagen, being welcomed by his two grand-children and a local guide, in time for a full historical tour. Thanks to a simultaneous translation tool, the guide explains – in perfect French – that he's currently physically in the Palace and that he's very happy to share a bit of his culture with Lauv and his grandchildren.

Once the visit is over, Lauv leaves his grandchildren in the capable hands of their Mandarin teacher. It's already 3 pm and he's not going to miss his weekly cooking session with his daughter Marie. From her kitchen in Dubai, Marie tells Lauv that today they will be preparing chocolate mousse. Marie has compiled a list of ingredients and she and Lauv enter the virtual Amazon Kitchen Store. They wander the aisles like they used to do in the supermarket when Marie was a kid. As they shop virtually, promotions and suggestions appear. Orders are placed for dark chocolate, sugar, eggs and all the other ingredients. A few minutes later, they are simultaneously delivered in Lyon and Dubai by drone, and Marie explains the recipe – and her secrets for making it – to her father.

An hour later, Lauv's grandchildren return from the Mandarin class, right on time for the tasting. It's a shared moment between the generations that crosses all borders. It's also an opportunity for Marie to greet her nephews and to enjoy the dessert together, at least virtually. All that physically binds them together is the smell of the chocolate transmitted by sensors on the 360 glasses. Yet, this virtual reality allows Lauv, his daughter and his grandchildren to spend quality time together far more often than they did before.

REDEFINING BOUNDARIES AND OFFERING A NEW WORLD OF POSSIBILITIES

After the Covid pandemic of 2020, companies changed forever. For many, brick-and-mortar offices became a thing of the past. Up until the end of the 2020s, tools were continuously improved to offer an optimised, at-home work experience. Company buildings gradually became obsolete. Most were demolished or re-purposed to benefit CSR projects. Even a giant such as Lauv Ltd was no exception. It began with all the furniture being given away – some initially to employees, and then to charity organisations. Eventually, by 2029, the office building in the centre of Lyon was pulled down to make way for a public garden. All over the world, it was the same. Although Apple kept Cupertino's emblematic building intact, others were not so lucky. Those few that remained, were turned into state-of-the-art co-working spaces for workers whose homes were too small for a functional office.

Each company needed to devise its own policy regarding working at home. To help simplify the new working landscape, employees were increasingly asked to keep their office rooms minimalistic. White or pastel coloured walls became the norm. Furniture was kept simple, just a desk with a spot to recharge 360 glasses and headset, an ergonomic seat and a coffee table. Cameras, installed by the employer, were placed in each corner of the room, to allow a realistic, 3D rendering of the employees during virtual meetings.

Work was just the first step. Soon, virtual presence became commonplace. Let's say you happened to be living in Seoul, and you wanted to go on a quick tour of Paris. A couple of steps into your exploration room, and *voila*! Thanks to Deepl's simultaneous translation tool, you'd understand a local guide perfectly. A weekly meeting with siblings on different continents? No problem, you'd just pair your 360 glasses together and meet for a walk in the park you used to go to when you were kids. Your exploration room would be able to reproduce the place down to the last detail – even the sweet smell of cut grass – thanks to the fragrance sensor on your 360 glasses.

By 2051, you could pretty much do everything from home. Cook with your family, purchase groceries and walk the aisles of a supermarket, play games with your friends, visit your family and of course, work with your colleagues. Virtual had become the new normal. For Lauv, it was all he needed. Except perhaps, a cuddle from his grandchildren.

REAL OR VIRTUAL? TOWARDS A NEW UNDERSTANDING OF THE WORD 'PRESENCE'

The phone has enabled us to talk to each other remotely. The smartphone has added our faces to our voices so we can 'see' each other from afar. The next step is not only known, but perfectly realistic. In May 2021, Google presented Starline, a mirror-screen that enables face-to-face conversations with depth perception. This is the first step in an evolution that will allow humans to be virtually in the same place. Until now, the technology has been used in television to conduct duplex interviews. Face-to-face will then give way to the notion of physical presence in the same place (real or virtual), thanks to the evolution of connected glasses and headsets, Google Glasses having pioneered this concept.

The 2020 lockdowns introduced us to social distancing for the first time in contemporary history. This phenomenon has led us to explore the demand for remote collaboration tools, where headsets (AR and VR) will play a major role in the coming years. Facebook has accelerated the development of productivity applications for its Oculus for Business headset. Two solutions have been introduced: Spatial and Immersed. The first allows any room to be transformed into an immersive workspace, where each participant appears through their avatar. Immersed is more of a concentration space: it connects people and virtually teleports them to the same space so that they can interact.

These headsets will present us with a whole new range of opportunities. Just take virtual supermarkets, for example. In 2016, Keiichi Matsuda shot a visionary video featuring augmented reality elements in a very real supermarket. Two years later, Amazon unveiled the first draft of a fully virtual supermarket at Prime Day in India. The aisles of the mall can be visited in an enclosed space, where you can literally walk through the stalls of each brand using a simple VR headset.

With a sophisticated headset, virtual travelling doesn't seem farfetched. Ava Robotics, among others, has developed intelligent telepresence robots to ensure virtual teleportation. This has implications for the professional world – as Owl Pro does for meetings. Similarly, in the fields of culture and tourism, schools and universities, players such as Double Robotics are already present. While virtual reality currently has room for improvement regarding image definition and realism, thanks to increased computing power and speed, these obstacles will be removed in the decades to come. Soon, we will see speeds being expressed in hundreds of gigabytes or even terabytes per second. The first hundreds of gigabyte connections are expected in early 2030. And by 2051, 8G is likely to be installed in most populated areas of the world, allowing tens of terabytes per second to be passed.

The only unknown is how our smart cities of the future will look. The notion of hybrid work introduced after the pandemic should accelerate the reduction of office space, that can then be replaced by green spaces and gardens. Buildings could become virtual. So too could architecture, which means that office real estate will need to be reinvented for a new era. This step is the logical continuation of industrial augmented reality applications imagined by Microsoft with Hololens in 2017.

PROBABILITY

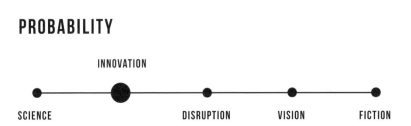

When we consider how – in the space of just a year – a pandemic could so drastically change our habits, working with a headset is just around the corner. There's no doubt that the technological improvements of the past few years will continue apace. Our future lies in virtual reality and it's not insane to think that this will intensify. In addition to the Oculus headsets mentioned previously, Facebook has also recently launched its own virtual workrooms for better collaboration. These workrooms might soon become the future of our current meeting rooms – working with VR headsets. Using spatial audio, they'll give participants the feeling of being in the same room as others. And there's more; the virtual room can be configured according to the needs of the meeting – be it for a presentation, collaboration or conversation. New features will soon be developed, allowing users to have their own avatars and mixed-reality desks. If this is the future of work, it's not difficult to imagine the same outcome for play.
Welcome to the virtual future.

ROBOCOP 2.0

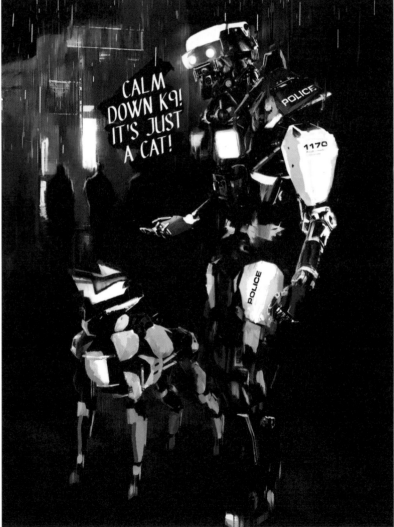

Belgian Fashion Week, 2051. Luigi taps on his smart glasses to check his team's progress. His view of the Bruges cityscape – now mostly underwater – dims as he sees their estimated arrival time: 11.07 am. This is right before historical house Leyson will dispatch their brand-new collection of healing jewellery pieces to be shown in the 2051 Valentino Fall ready-to-wear collection. These exquisite ear, forehead and skull cuffs are made from the super-rare metal iridium and fitted with ultrasound emitters that can treat migraines and anxiety.

Luigi's crime ring has had their eyes on these innovative jewels for a while, and today they're going to smartly intercept them. Just as the high-security delivery drone leaves the maison, a mesh of harpooning drones will attack it from all sides and pull it onto the rooftop of a luxury hotel near the Bruges City Hall. Here, a robodog will force open the drone chamber with just the right blast of explosives, grab the package and fly off to a secret location in the Port of Zeebrugge. Luigi is watching all the action from an apartment opposite the hotel.

Little does he know that Interpol has been all over this little operation for months now. Working with Belgian police, it has used sound sensors attached to buildings to monitor unusual drone sounds. Dubbed the 'Maltese Mob', this crime ring is known for using audio camouflaging with their attack drones when it's metres away from the target, to try and confuse audio sensors. The sound? The hysterical barking of Maltese poodles. This annoying sequence has been traced from Milan to Paris Fashion Week. It was only a matter of time before they would strike in Bruges. And the teams are ready and waiting.

The moment the sensors pick up the tell-tale barking sound, autonomous patrol boats start chasing through the canals, following the scent like bloodhounds. They have the smartest ethical hackers on the team to trace connections to the crime drone, and quickly pinpoint Luigi's location. Patrollers quietly enter the apartment building, slip upstairs and break down his door to arrest him...just in time to stop the robodog from exploding the shiny, high-security delivery pod. Luigi is marched downstairs and into the awaiting police boat, which is kitted out to be a mobile charge station and interim courtroom. His iris is scanned to confirm his identity and he is covered with a body sensor 'polygraph jacket'. On the boat's screen, he's shown images of the stolen collection. *From his reaction – dilating pupils, increased pulse and rapid breathing – it's clear that he's guilty of masterminding.* 'You have the right to remain silent', says a hologram from the dashboard, as it reads him his rights and auto-dials in a hologram attorney.

YES, YOUR VIRTUAL HONOUR?

As we speed towards 2050, government bodies in charge of law enforcement will need new ways to cut costs. First up will be 'smart courts': AI-powered hologram judges who try cases during digital court hearings. Defendants, witnesses, victims, lawyers and the jury will attend court as holograms. Technologies like holograms will make victims more likely to come forward, as they no longer have to face assailants in person if the case goes to court. Those who do decide to attend the courtroom in person will be greeted by a smart court robot, guiding them to their appropriate seating locations. Stenographers will be a thing of the past as court hearings are now fully recorded and searchable via AI-generated scripts. Everyone with access to court documents will view hearings in real-time from the comfort of their homes, with an AR courtroom and participants.

SENSING CRIMES BEFORE THEY HAPPEN

Parabolic microphones will still be widely used to help detect sounds far beyond the normal range of human hearing. But the way sound is used to policing advantage will expand even further. Gunshot detection via sound detecting sensors will become the main tool for police department collaboration, knowledge exchange and data collection for deep learning. Sound sensors will be placed on the corners of buildings around neighbourhoods. If shots are fired, they automatically detect, record and timestamp the sound. Besides the sound location, AI will also determine which firearm was used based on the sound database. Data is then sent to the nearest police department, and a scene inspection task is delegated to officers who are patrolling that neighbourhood. Deep learning of gunshot data, crime locations and popular times of the day will enable effective city mapping to predict future crimes. AI neighbourhood police cameras will detect unpredicted crimes as they'll be able to distinguish criminal activities from basic ones. Data on the type of crime committed will be automatically shared with the patrol car nearest to where the camera detected the activity. Remember those exhilarating high-speed chases you always saw in Hollywood movies? In the US, these out-of-control police pursuits killed many innocent people. This is why GPS darts will make a comeback in the 2030s, when the problem of them not sticking on wet and cold surfaces will be resolved. Darts ejected from police cars glue themselves onto the surface of the pursued vehicle to slow it down. Cop cars will get a revamp too, with driverless vehicles becoming a common sight. Speed cameras and sensors can't possibly catch everyone driving above the speed limit, therefore autonomous police cars with built-in speed scanning systems will start to patrol the streets. As they drive around, they would measure vehicle speed and automatically issue tickets. Police departments will get another powerful back-up in the shape of robot police officers and dogs – especially in high-risk situations where bombs and hostages are involved. The robots will have Range R radars installed, allowing them to detect motion through solid walls.

Crime reporting will be much more accessible via 3D chat booths. One police-man will remotely oversee several booths in multiple neighbourhoods where citizens can physically come to report crimes in real-time, making everything much faster and more efficient. The traditional 3D crime-scene imaging method will be elevated with holograms, allowing you to see holographic crime scenes and event reconstructions. Lie detection like polygraphs will be failproof and court admissible. In previous decades, cheap computing power, brain-scanning technologies and AI gave birth to a powerful new generation of lie-detection tools and interrogation techniques – e.g. memory detection. This works by seeing whether a person has guilty knowledge of the criminal event based on his reactions, heart rate, etc.

CATCH ME IF YOU CAN

Smart courts, 3D crime scenes and GPS darts. Do some of these criminology predictions seem a bit far-fetched? Not if you start digging around. China has been using AI judges and courtroom holograms for many years. In 2019, it announced that millions of legal cases will be tried by 'internet courts' that don't require citizens to be physically present. Estonia is another country planning to implement a 'robot judge' for disputes of less than 7000 euros. As for stenographers, it won't be long until governments realise how easily costs can be cut by replacing them with AI. Gunshot detection is in use, but not by many police departments. ShotSpotter is gunshot detection techno-logy that uses sophisticated acoustic sensors to detect, locate and alert law enforcement agencies and security personnel about illegal gunfire incidents in real-time – less than a minute after shots have been fired.

AI could also make its way to neighbourhood cameras. In 2021, Facebook announced Ego4D, a project aimed at teaching AI systems to comprehend and interact with the world from a first-person perspective – just like humans. The AI will be fed sufficient data to help them group activities together (e.g. sports and shopping). It's easy to imagine that soon AI will be able to distinguish criminal from harmless activities. UK police are already taking advantage of predictive crime mapping using the PredPol system. Following a report stating that British police have a wealth of data but lack the capability to use it, they decided to pinpoint high-crime areas, and times and days of the year when crimes take place. The type of facial recognition of criminals we mentioned is far less intrusive than border and airport cameras. Software like Clearview AI is already in use by police forces around the world. It collects billions of images from Instagram, Facebook, Twitter and YouTube, without permission, then links images from surveillance cameras to an identity or profile on social media. GPS darts present somewhat of a solution to high speed car chases but are yet to be widely enforced. They're only useful under special weather conditions when the glue can actually stick to the frame of the car, so hopefully there will be an invention soon to make darts weatherproof.

Autonomous vehicles are a hot topic, but there hasn't been much talk of an automated police car. In 2018, Motorola filed a patent for a 'mobile law enforcement communication system' that would be a law enforcer and court room on wheels. The car would feature a live communications functionality that would allow judges to question suspects via a video feed. Though we still don't have robotic policemen, Spot, the agile mobile robodog, is already finding its way into police departments. In 2020, NYPD police used the dog to find a gunman who'd barricaded himself in a building after he'd accidentally shot someone in the head. Police departments all over the world are increasingly welcoming the idea of robots. Dubai, for example, wants to become the world's safest big city. In 2021, Dubai Police signed an agreement with UiPath – a leading global software company – to enhance cooperation and collaboration in robotic process automation (RPA). Dubai hopes that robots will constitute 25% of its police force by 2030, which includes using them as receptionists in police stations.

3D scene imaging has revolutionised crime scene investigation. Holograms are yet to receive their crime case application we saw in movies like *Kin* and *Iron Man*. We could imagine crime scenes reconstructed in VR as well, but we'll choose holograms for the convenience of being able to collaborate with multiple experts at the same time – without the need of bulky VR devices.

Should polygraphs be admissible in court? The debate continues. However, there's no question that people are really bad at identifying lies. Recent research shows that polygraphs are far more accurate when done together with Concealed Information Tests (CITs). Thus far, Japan is the only country where a polygraph and CIT combo is widely applied to criminal investigations. Personal genomics and biotechnology companies have already helped to solve a significant number of cold cases. In 2019, a woman shared her DNA with the 23AndMe database to discover more about her heritage. After some research, she found out that she could further share her results with GEDmatch, a free open-source DNA database accessible to law enforcement without a court order. GEDmatch then confirmed her DNA matched a suspect at a third-cousin level, who was then successfully charged and arrested for the 1980 kidnapping, rape and murder of Helen Pruszynski.

PROBABILITY

Though many crime fighting technologies already exist, who knows how soon we'll see these things on our own street corners? Implementation always comes down to costs and most tech companies are more interested in innovating products than streamlining production processes. Spot the robodog, for example, has a hefty price tag of $75 000. It's hard to imagine police departments willing to risk blowing it up during a bomb deactivation process. Perhaps Chinese start-up Weilan will get us closer to reality. Their AlphaDog comes at a significantly lower price of $2500.

The tech we'll definitely see implemented are holograms, AI-powered judges, AI stenographers and 3D projections for crime scene investigation. We'll also see the rise of advanced law enforcement data-collecting platforms to solve crimes with deep learning and AI.

WELCOME
TO THE SPACECATION

Since her holiday began, Lula hasn't missed a single sunrise. Each of them has been successively more beautiful than the one before. The tenth one takes her breath away – the sun's rays break over the horizon just as the Aurora Borealis spills its last green and purple ripples across the frozen wastes of northern Finland. The light is blinding. She blinks against the glare and drifts away from the window, steadying herself with her right hand on its sturdy, grabbable frame. Her left hand is comfortably at eye level and so is her gyrowatch. Could it really be 1.30 am? Time for bed. She glances upwards as she makes her way towards the exit. Friends Wes and Kai are leaving too. Wes uses the highest handles to propel himself down into a backflip, while Kai follows more slowly, crossing the ceiling at a steady pace and then walking down the wall to the floor. It's the first time she's seen either of them since they all first checked in together at the docking station this morning. What a day. She feels her heart rate fluttering just a little. Could it just be the low-g? Or maybe Wes is actually not bad looking.

A minute later – and a couple of graceful glides down the hotel banisters – Lula is back in her room. There's much more gravity in here than in the viewing pods – the science of which she fully understands, but still can't quite grasp in reality. The space hotel she's on – the HelixXX777 – was only recently completed, its components having been built entirely in space. She wishes the new building currently going up next door to her apartment block back down on Earth could be similarly constructed. It's only 250 floors but the noise levels have been through the roof for months now. At least up here, it's nice and quiet.

She sits briefly on the sumptuous hotel bed and smiles. In the early 2020s, after the first Covid outbreak, back when the world had been in a thing called lockdown and online schooling was new and very boring, she'd watched endless videos about the day-to-day lives of astronauts at the ISS. It had been a great distraction. She's happy to see that, unlike them, she doesn't have to zip herself into a sleeping bag attached to the wall and sleep upright. Instead, there's a duvet on the bed. It's goose-down, like the ones her mother used to sleep under. Now that technological advances in farming have made it possible to harvest feathers painlessly from the birds, animal fibres are finding their way back – even if initially in only the most exclusive establishments. And with temperatures up here being markedly chillier than down on Earth, Lula is grateful. Lula voice-commands the blinds to close and the lights to dim, as she makes her preparations for bed. As beautiful as they might be, she doesn't want to be woken by the next five sunrises as she tries to cram in her eight hours of sleep before breakfast. The blinds have been programmed to lift just before the sixth sunrise, so she will be able to watch it from her bed. According to the chatty little InfoBot she'd encountered in the gravi-elevator earlier, the sunrise she has in mind should incorporate gorgeous views of the Amazon Basin. It's expanded by a third over the past five years. She doesn't want to miss this satellite's eye view. Walking into the en-suite bathroom in her memory-wrap robe, she sees the 360-degree rainforest shower. On entry, she is automatically doused from all angles with very fine jets of water. Every drop will be retained and reused on the space hotel. Soap (to use the traditional term) is pre-programmed to target just the right areas. No more nicking sample size hotel toiletries. Beyond that, she also knows this shower is a source of vital data collection via monitors on the shower walls. Her bone density will be scanned and measured every day of her spacecation. So too will the water levels in her body. The pH levels on her skin will be sampled and analysed. As well as, of course, body weight fluctuations, muscle mass and lung capacity. It's become the quid pro quo of space travel. A significantly discounted holiday in return for personal info. Of course, there are those who say it's a violation of personal privacy. Just like those who, back in the days when Covid passports were first introduced, had maintained that governments and corporates had no business regulating travel with a vaccine score card.

But whatever qualms she had about submitting herself to these rigorous – some would say slightly invasive – tests, she feels it's worth it. All the data collected from her short visit to space will be used to accurately inform future, long-haul space travel. The first long-term mission to Mars is departing from the Moon in three months' time. After a decade of short trips to the Red Planet, this will be the first attempt at a twelve-month stint. And now that her brother Abe has just been selected to be a part of that mission, it just came closer to home. So, she's happy to do her bit for science – even if it involves standing naked in a shower on a low Earth orbit journey.

The next morning finds Lula in the breakfast room, watching the sun setting over the Pacific. A nervous looking WaiterBot serves her up a legume and argonut waffle with a molecular blueberry foam and some super-hydroponic strawberries – the first ones to be cultivated in space. They're small but taste delicious. She snaps a quick pic on her gyrowatch to send to her mother. Even though it's 2051, Mum still nags them all about eating their five-a-day. She can hear some muffled sounds coming from the kitchen. She laughs to herself. Hiring the classic GordonRamseyBot as head chef was an inspired move on the part of hotel management, but it doesn't come without its challenges. No wonder the poor WaiterBot's screen is glitching slightly.

She puts on her AR glasses to check the schedule for the day. Low-g basketball. Tick. Freeze-dried macaroon making classes. Pass. An afternoon spacewalk. Double tick. The Twenty-Fifth Anniversary of Space Travel Dinner with Dancing? Not missing that for the solar system. Good thing she thought to pack a party dress with a specially weighted hem. She secretly hopes Wes and his wingman will be there. What was that song Grandpa used to sing again? Something about Dancing on the Ceiling. She smiles. Only another twelve sunsets before it's party time...

THE (BILLIONAIRE) HITCHHIKER'S GUIDE TO THE GALAXY

For most of us, travel during 2020 and 2021 was a simple affair. A well-worn route between the kitchen and the sofa. At least there was no visa required. So perhaps it wasn't surprising that the news of Sir Richard Branson and Jeff Bezos's respective trips to space was greeted with such fanfare. However, it's easy to forget that they weren't the first billionaires in space. Whilst Blue Origin and Virgin Galactic have been lauded for their recent achievements, Washington-based Space Adventures have, for the past twenty years, been organising space trips for privateers with a private equity sized fund to burn in jet fuel. Notable examples include Mark Shuttleworth, who became the first African in space in 2002, Dennis Tito, first privateer (2001) and Anousheh Ansari, first Muslim woman (2006). Working in tandem with Russian-based Roscosmos, these lucky few have had the opportunity to hitch a ride on a government-owned rocket and orbit Earth.

So, what's changed? Unlike earlier space tourists, Bezos and Branson – along with Elon Musk – have made no secret of their mutual ambition to commercialise space travel. But who of us has the means to make that possible? And why does it matter? It's tempting to dismiss these latest trips into space as mere flights of billionaire fantasy. But these disruptors – and the rivalry their efforts have created – have opened up a new frontier for space travel by developing their own spacecraft. And competition is always good. According to NASA, it's resulted in the cost of reaching low Earth orbit being reduced by a factor of twenty, when compared to the conventional, government-funded model. Cost reduction has extended to other areas, too. Cheaper launch costs, reusable rockets and an increase in the number of satellites, all make for an exciting, and hopefully increasingly more affordable, space race of the future. Behold the era of NewSpace.

But does it all add up? It's hard to forecast, but back in 2017, Morgan Stanley and Bank of America Merrill Lynch predicted exponential growth for the space industry in the coming decades, reaching a market value of $1.1 trillion by 2040 and almost $3 trillion by 2050. With the increase in investment, and plummeting costs continuing, space will become increasingly more accessible.

FLY ME TO THE MOON

What will the spacecraft of the future look like for future galactic tourists? The best indication of this may just be found in NASA's Artemis Programme – the successor to the Apollo missions of the 1960s. Artemis has committed to sending the next man – and woman – to the Moon by 2024. For this latest voyage, NASA has enlisted the help of private companies – including SpaceX and Blue Origin – to develop a suitably out-of-this-world spacecraft which they've called Orion. Designed as a space capsule, Orion will be able to accommodate a four-person crew. This signals a move away from the shuttle, which over the years has had some tragic failures. Also, unlike the shuttle, the capsule can fly beyond Earth's orbit, its design is better suited for a lunar landing, and its launch-abort capabilities offer better protection against rocket malfunction. The capsule is also lightweight enough for long-haul journeys – possibly to Mars. Blue Origin is currently developing its own lunar lander, known as Blue Moon, capable of sending crew on a Moon mission. And if it's good enough for the next moonwalk or a mission to Mars, it should be good enough for well-heeled Gen Beta travellers.

Virgin Galactic has already revealed the latest design for its next generation Spaceship III prototype. With a mirror-finish cladding, the VSS Imagine's sleek shape and super-reflective surface is a powerful symbol of the transformative power of galactic travel. Like Unity, the Imagine will be launched from a mothership, called WhiteKnightTwo, where its rocket engine will propel it into sub-orbital space. Could this be enough to take a craft further, and possibly out of Earth's orbit? Who knows. But already, the company is now preparing to fly 400 sub-orbital missions a year from each of its spaceports. This, after almost twenty years in the making. (Spacecrafts, it seems, are a bit like London buses. You wait ages for one, and then all of a sudden, several appear).

Whatever the future may hold for spacecraft design, space tourism companies will be held to the highest levels of safety scrutiny. According to the UN Office for Outer Space Affairs director, Simonetta di Poppo, issues such as these have already been given consideration at a recent UN Committee meeting on the Peaceful Uses of Outer Space (COPUOS). All operators will be obliged to obtain a licence from their authorities. And in accordance with the 'Outer Space Treaty of 1967', no single entity or body operating in space will be able to lay claim to land on the Moon or anywhere in space. The 'Moon Agreement', ratified in 1984, provides that the Moon and its natural resources 'are the common heritage of mankind'. The Agreement also states that an international regime should be established to govern the exploitation of the Moon's natural resources, should that become feasible. All of which means that, with a bit of luck, and some new statute books, we can rest assured that the final frontier won't become the new Wild West.

POSTCARDS FROM THE EDGE OF THE SOLAR SYSTEM

If the first port of call is to be the Moon, the next stop will be Mars. It's predicted that between the 2030s and 2060s, astronauts from no fewer than three major space agencies will be making the six-to-nine-month journey to the Red Planet, with NASA's Mars to Moon programme expected to be at the forefront of this exploration. More nations will become members of the official 'space club'. Earth's orbital lanes will become busier, with the commercialisation of the low Earth orbit and an increasing number of satellites entering space. And with the ISS due to retire in 2024, successive space stations will surely fill this gap.

In the coming decades, these will make way for more advanced stations, developed under the banner of the Gateway Foundation. Among these will be the Lunar Gateway – a small space station situated within the lunar orbit, serving as a halfway house to deep space. The role of the Lunar Gateway – and likely other space stations – will be that of providing testing centres for advanced technologies that could support life in space. These galactic outposts will be designed and constructed along the lines of a rotating pinwheel, using 3D printers and autonomous robotic arms. The construction of these new stations will be overseen by the Orbital Assembly Corporation (OAC) – the first large-scale orbital construction company in the world – or should we say, solar system. Their flagship project is the Voyager-class space station. Built to be a giant, spinning ring, it will be capable of simulating gravity. The OAC's stated mission is that of providing 'the comfort of low gravity within the luxurious accommodations of a luxury hotel space for tourists who want to experience an extended visit to space'.

Of course, simulating gravity will be the potential game changer for the future of space missions and, consequently, of space travel. Should this become a reality, we'd be one step closer to sustainable life in space, certainly for an initial period of up to 24 months. For now, the focus looks set to be on the establishment of low-orbit hotels. One possible design is that of a hydrogen-filled balloon with a pressurised capsule, which utilises earth's gravity. Fellow space operator Bigelow has developed a design for an inflatable habitat, which it plans to attach to the ISS as one of the first hotels in space.

Another audacious plan is that of the Aurora Station, developed by Orion Span, which can accommodate six guests for a $9.5m, twelve-day luxury vacation in low Earth orbit. Flying 200 miles high, the Aurora will orbit Earth every 90 minutes, offering sixteen sunrises and sunsets over a 24 hour period. Luxury sleeping pods, gourmet cuisine, low-gravity ping-pong and the experience of growing food onboard, will all be included in the hefty price tag.

PROBABILITY

INNOVATION

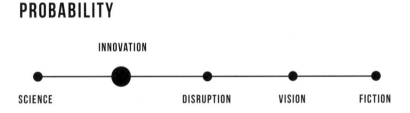

SCIENCE   DISRUPTION   VISION   FICTION

It is overwhelmingly likely that, by 2050, space tourism will have been firmly established as a small but significant subsector of the NewSpace industry. Low Earth orbit long weekends, lunar holidays, perhaps even long-term Mars-batticals – all these could become a reality, with the strengthened and mutual resolve of governments and privateers. Job creation – be it in the sector of space aviation, orbital construction, or solar-electric propulsion – all these will vastly alter our prospects for development well into the 21st century.

Above all, surely witnessing the awe-inspiring beauty of our planet from space, even if only for the initial fortunate few, should be the lightning rod that alerts humanity as a whole to the urgent task of forging a sustainable future within our solar system – and beyond.

FACES
FOR SALE

Fatimah is working the room at the fundraiser, feeling absolutely fabulous. She knows that her focus should be on raising money – a whole lot of it – to help restore a habitat for wild tuna. Since the beginning of the century, the delicacy is no longer available, thanks to trawlers having completely destroyed their eco-system. Instead, Fatimah picks up a dried seaweed cracker from a tray carried around by waiterbots, knowing it won't interfere with the make-up on her lips.

She wants to keep her 'face' intact – after all, she's paid enough for it. Hoping everyone will notice she's wearing an original De Backer tonight; she smiles at guests and laughs a little too loudly. And they all look at her. Why wouldn't they? The artist's signature style is instantly recognisable – coveted by art lovers since it was sold as the most expensive look in the history of make-up NFTs. That's not the one she's got printed on now (her husband's taste sadly wasn't that on-point), but it's from the same collection.

A special twenty-five-year anniversary gift, he made sure it came with a face printer and the exact colour cartridges to match the resolution that the artist intended on-screen. Fatimah has been following a strict skincare routine and eating plan for over a month to get the most flawless printing done for this evening. Her face is perfectly smooth, her pores small, with not a single oily patch in sight. She's been scanning her skin religiously in her smart mirrors to get a personalised skincare routine for every second of the day. She even carried a miniature skincare dispenser in her purse to get the exact amount of personalised product whenever her skin condition changed. And it worked marvellously. She's turned her skin into a perfect canvas.

Looking around, Fatimah notices how few smart glasses are worn on this special occasion. A lot of people have pulled out the stops to look their best tonight, trading eccentric digital make-up for ephemeral, skin-printed art that must've cost a small fortune. Some opted for obscure, up-and-coming digital artists, who are hoping that showing off their work at this offline benefit tonight will lift their value online. Not Fatimah. She only prints the big names from her blockchain make-up collection for real life events. She's been doing fundraisers to restore extinct species for more than a decade. She learned early on *that she needs to exude wealth, luxury and good taste. Tonight, she needs her guests to feel jealous so they will throw ridiculous amounts of money at the good cause to upstage her.*

It's not that Fatimah doesn't support the digital make-up scene – she owns a small but interesting collection of young artists. But she keeps the lesser-known works to wear on her digital face, when attending online meetings or going to lunch with potential donors. Avatars in the metaverse don't require flawless, real-life skin.

WHEN MAKE-UP MEETS BLOCKCHAIN

Make-up artists, even the famous ones, don't produce work that increases in value. Yes, the more famous make-up artists and influencers charge a higher rate for their time, but the work itself can't be traded. Their design is applied onto a face, lives for a day, and disappears again within 24 hours. An Instagram picture captures their work, but that too is not a tradeable asset. We predict that all this will change in the late 2020s, when many developments will converge.

The first trend is the stellar rise of NFTs as an art form. NFTs stand for non-fungible tokens – pieces of digital art that are encoded in blockchain technology. Because they're encoded, it's forever possible to identify the original digital version – no matter how many copies are made. This makes digital art collectible, just like offline art. Think of it as lithography. No matter how many prints are made, it's always possible to identify the original litho. With NFTs it's the same: no matter how many copies exist, the owner of the original will always be able to claim ownership of the one 'true' version of the art piece.

We predict that NFTs as an art form will reach a tipping point and become mainstream long before 2051. The reason for this is simple: they will generate income to rival that of a traditional artist. Therefore, registering digital artwork on a blockchain will become simple, easy and affordable for any artist.

Art collectors will flock to NFT marketplaces which will function like art galleries. Art lovers will collect digital works from their favourite artists.

But why would make-up artists go digital? Their work is inherently offline, so who would spend a few thousand euros on a look they can't even wear? We predict that the algorithms used for funny AR filters on social media will become open-source and platform independent, allowing anyone to build a face filter. As the algorithms improve, they'll become better and better at mapping human faces, making the filters extremely realistic. The projected 'face' will seamlessly cover your every move. Creating these lifelike filters will become super-easy, so it will only be a matter of time before make-up influencers start to create digital looks that anyone can wear. Using blockchain technology to monetise those looks will be a logical next step. Fans will be able to buy a specific look as if they were buying a piece of art, and then show it off at a digital meeting or on their avatars while walking around the metaverse.

In the 2020s, most self-respecting beauty companies are already heavily invested in beauty tech. They will begin to market smart appliances like mirrors and brushes to complement skincare and beauty products. Smartphone apps will scan a person's skin, analyse the input and create personalised skincare routines. This has already resulted in made-for-you beauty products based on your very own user data. We believe that this trend will continue, with personalised skincare becoming the norm – first for beauty influencers, then for followers. Generic products and one-size-fits-all skincare routines will disappear, replaced by intelligent beauty devices that offer instant, bespoke solutions. This hyper-personal skincare trend will be fuelled by the desire to have a machine-readable face so soft and smooth that any AR filter of choice will fit it perfectly.

There's another development worth mentioning here. Since the Covid pandemic, people have even less patience than before. When we order online, same-day delivery is non-negotiable. This desire for an instant service will filter down to other aspects of our lives. We'll also want our beauty products ready the moment our skin starts to feel a bit dry or blotchy. When there is no time for a beauty company to produce and ship a single product to you in an instant, they'll bring you another solution out of necessity: the make-it-yourself kind. Beauty tech will invest in hardware that lets you create the exact product you need when your smart device tells you. Using cartridges filled with cosmetic ingredients, you'll be able to mix products yourself. And, just like computers, the devices that are big and clunky today will be transformed into wearable, pocket-sized devices you can take anywhere. The result? Instant, hyper-personalised skincare to achieve the perfect skin canvas for your digital make-up. But why stop there?

There's one last trend we believe will add to the rise of the digital make-up artist: printing. Yes, you read that right: printing. Traditional art is hung on a wall and its owners wait for people to visit and admire it in person. But some art is mobile. Think high-end jewellery and watches. They can be shown off anywhere, anytime. Owners of expensive make-up NFTs may want to show off their high-end art the same way. To this end, the printers now used as in-store gimmicks to print nail polish on fingernails or eye shadow on eyelids, will evolve. They will become small home devices that will allow owners to print NFT looks on their own faces for a day and display them for the world to see. If they follow a personalised skincare routine to turn their faces into the perfect canvas, they'll be guaranteed the most flawless printing result.

A BIT FAR-FETCHED? THE WORLD BEGS TO DIFFER

If you too want to be served by a waiterbot like our dear protagonist Fatimah, you could visit the Royal Palace in Renessee, Netherlands, where two lovely robots in waiter outfits will do their best to make your visit a memorable one. It won't be long until we invite these bots into our own homes.

Print-on make-up? Highly probable. You might be a bit hesitant about printing something on an irregular surface like the human face, but there are printers that have already conquered this challenge. PrinCube, the world's smallest mobile colour printer, transfers ink onto almost any surface. Its creators vouch for skin as one of these surfaces and claim that their tiny printer is perfectly suitable for safe, temporary tattoos on skin.

Though there are some smart mirrors on the market, people are increasingly making them themselves by combining two-way mirrors, LED TV screens and Raspberry Pi computers, and transforming them with the help of the Magic Mirror open-source, modular smart mirror platform. It's only a matter of time before smart mirrors start dispensing personalised skincare and make-up guides.

An example of a dispensing device that could easily be integrated is the L'Oréal Perso. It contains cartridges with make-up and skincare which are mixed according to users' skin needs and look preferences. The analysis itself is done by a camera and sensor on the user's smartphone, which is then sent via an app to the Perso. Who knows if physical make-up will continue to be superior to digital make-up? It's likely that the two will co-exist and complement each other. Who hasn't been tempted by Zoom Studio Effect filters during Covid lockdown? Digital make-up filters could become more popular than you think. For example, dermatologists could recommend them to let skin rest and breathe after certain procedures.

Your digital or virtual presence will eventually matter just as much as your physical one. Some people are already ready for the transition to a virtual world, as they continue to buy pieces of earth on the Earth2.io metaverse. The next logical step is to start customising their lives in their digital world. Here, it doesn't matter that you have acne scars, rosacea or a nose that's a bit too big.

Printed make-up looks could become social status symbols too. This shouldn't come as a surprise, especially if we think back to games like Stardoll. This game has been in existence since 2004 and caters to an often-overlooked audience in the gaming industry – teenage girls and young women. It allows users to customise their dolls, buy them designer clothes and make-up. They use these things to show off and enhance their status. Clothing and make-up trends would go out of stock and change, just like in real life. On some highly regarded user profiles, you might come across vintage DKNY pieces that Stardoll released back in 2007.

The last thing that ties all the predictions together are NFTs. Their future might seem uncertain currently, but looks more promising when seen through a long-term lens. Users can already purchase 3D printer templates of clothing that they can print themselves once they purchase the NFT. Perhaps NFT make-up is not that crazy an idea.

The only concern with NFTs are the unsustainable miner fees and the ways in which mining is conducted. However, the Proof-of-Work consensus on which most cryptocurrencies operate will soon be replaced by the Proof-of-Stake, which is meant to drastically decrease the energy consumption needed for mining, making NFTs a much more sustainable endeavour. Ethereum has already committed to adopting this consensus.

PROBABILITY

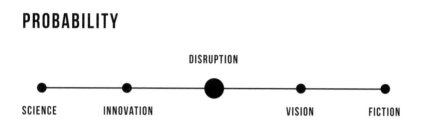

DISRUPTION

SCIENCE   INNOVATION   VISION   FICTION

There are already so many technologies supporting the futuristic concepts of digital and physical make-up. Art is becoming more valuable than ever before, and artists are very adamant about not letting what they do be dismissed as any less of an art piece. The world of NFTs has opened so many doors for them, giving them an opportunity to authenticate and trace the origins of their creations.

On top of that, artists will soon be able to detect whether their NFT is being copied and sold by someone else using duplicate detection by DeviantArt, the social network art community founded in 2000.

Though this appreciation for different types of art, especially digital, is quite new and refreshing, the cosmetics industry has been around forever. It has consistently grown, with only the pandemic having temporarily dented its results. Make-up and cosmetics are still very much connected to point-of-sales, because users like to see and test textures and colours. The industry will certainly bounce back post-Covid, and will start wowing us with all these new inventions.

Beauty lovers will always check themselves out in store window reflections. In 2051, Fatimah will notice something completely different. As she passes her favourite drugstore, the window will come to life with her enhanced reflection – augmenting the latest make-up looks available onto her face. Whichever version she prefers – digital or print-on – it will be available right there on her own high street.

CLOUDY WITH A CHANCE OF ARTIFICIAL RAIN

January, 2051. The height of summer in Port Douglas, Australia. It's the weekend, but Noah is up early to enjoy some quiet solo time at the harbour before the tourists arrive. As he prepares his boat for his diving trip to the Low Isles, he gazes at the sky. A bright dot suddenly crosses the horizon and disappears as the sun rises. Noah feels lucky to catch a glimpse of the MirOne orbital satellite. Thanks to its sun-reflecting mirrors, summers in Port Douglas have become more bearable.

At 7 am, the sun's up and Noah is approaching the Low Isles. Thick, white clouds cover the area – the saline blasters must have visited during the night to cool the air. Noah drops anchor on a sandy seabed before reaching what he came to see here today: the corals. He puts on his wetsuit and jumps into the water. It is not warm today, barely 22°C. It's been a few years since the temperature was lowered, thanks to the saline blasters.

The underwater landscape is magnificent. The corals share the stage with a multitude of fish species, shellfish and sea turtles. Just 15 years ago, the days of biodiversity were numbered. Dropping water temperatures around the island has been a lifesaver. After several hours of admiring the beauty of the seabed, it's time to leave. Time goes by so quickly on a dive but it's noon already and Noah must get back for a lunch with friends.

The terrace of the Surf Club and Bar is crowded but luckily his friends have saved him a spot. The building, recently refurbished, is dazzling white. The state of Queensland recently passed several laws obliging residents to paint commercial buildings with a highly reflective, white coating. Noah likes the fact that it brings brightness to the city centre until quite late in the evening, and that it cools the air in summer. With government financial aid, he also painted the roof of his small house bright white to help cool him down during the summer heat. Lunch is in full swing, and Noah shows his friends the photos he took that morning of the reefs near the Low Isles. After two hours of eating and chatting, he finally goes home to rest. That afternoon, while lying on his lounger with a good book, a light cloud cover protects him from the sun's harmful rays. Also artificially created, these clouds give the sky a pretty, light blue colour.

Noah still remembers his first summers as a child, suffocating in the blazing heat. He's relieved that he can now enjoy a pleasant afternoon in the middle of January, when temperatures are around 26°C.

As the evening approaches, Noah decides to call his sister Ava, who is on an expedition with a team of glaciologists in New Zealand. Thanks to her team's efforts, the glacier has regrown in thickness, compared to previous years. They have just finished covering it with an enormous, highly reflective white sheet. Ava expects to return to Australia in the next few days to visit her brother.

After this family interlude, Noah heads off to Oak Beach to meet another group of friends for dinner. As night falls, he looks east and sees MirOne appear once again, even before the Shepherd's Star. Watching this beacon of hope, Noah feels grateful for the special times he can spend with friends without the extreme heat he experienced as a child

THE SHIELD OF NATURE

Earth's climate depends on two main factors: the solar radiation it receives and the way it manages this received energy. 71% of this energy is absorbed and then released in the form of infrared radiation. The CO_2 in the atmosphere captures some of this radiation and prevents it from escaping, causing temperatures to rise. This is the dreaded greenhouse effect and the reason why climate change is such a growing concern. In the summer of 2021, temperatures soared, reaching almost 50°C in Canada. It is more urgent than ever to act, and researchers are already turning to a potential solution: geoengineering. What if humanity could control the climate in 2051? Stopping the rise in temperature is non-negotiable for the survival of mankind. Research into geoengineering will quickly become a global project, likely in response to a catastrophic event that will force governments to act decisively (much like during a pandemic).

Every year, for one week, air-raids will take off to distribute aerosols into the stratosphere. Their route will follow the equator, spreading aerosols freely around both hemispheres, and creating a light, protective cloud cover capable of reducing solar radiation by 1%. Temperatures on Earth will then be better controlled, and summer heat spikes will become increasingly rare.

The effectiveness of this initiative will encourage researchers to turn to maritime stratocumulus clouds. Lightening these thick clouds will make it possible to create a protective screen against solar radiation, thanks to its high reflection capacity. How can this be done? By injecting salt particles from sea water directly into these clouds. This initiative will be applied locally to lower temperatures and prevent coral reef bleaching.

Governments will encourage roof renovation for homes and other buildings with high albedo materials (the measure of reflectivity of a material). New buildings will come with these types of roofing as standard, which will reflect the sun's rays in a more optimal way. In summer, glaciers will be covered with large reflective tarpaulins to preserve them, while largely shielding most of the sun's rays.

In parallel, efforts to absorb CO_2 emissions – released as a result of industrial activity – will be hugely ramped up. The cultivation of phytoplankton will be commonplace, absorbing up to 50% of these emissions by transforming them into sediment. Reforestation projects will be widely supported, including in some desert areas.

A coalition of researchers and engineers will work tirelessly on a project of interstellar proportions. It will be a reflective orbital satellite that will act as a giant, travelling mirror. It will reflect the sun's rays before they even enter the Earth's atmosphere.

These initiatives will buy humanity time to find a way to further reduce their CO_2 emissions and adapt to future growth. This is not a quick fix or an excuse for inaction. To avoid a terminal shock phase, we need a complete rethink of how economies and societies function. A phase in which, if geoengineering initiatives were to end for whatever reason, the effects of climate disruption would be felt even more quickly and strongly than before, spelling the end of humanity.

CONTROLLING EARTH'S THERMOSTAT

Pure science fiction? Not entirely. Leading scientists are already looking into the subject and many research studies have been done. Dr Anthony Jones, a scientist working on the Earth's atmosphere at the University of Exeter, has highlighted the fact that injecting aerosols into the stratosphere could be particularly effective. He draws parallels with the 1991 eruption of Mount Pinatubo in the Philippines, which was so violent that millions of sulphur dioxide particles were suspended in the atmosphere, creating a protective veil that caused mean global temperatures to cool by 0.5°C for several years.

In the race to cool our burning planet, it seems the old adage might be true – every cloud has a silver lining. Researchers Douglas G. MacMartin, Katharine

L. Ricke and David W. Keith have been working on the effects of injecting salt particles into marine stratocumulus clouds, which could help to cool local temperatures. In addition, studies by John Latham, Joan Kleypas, Rachel Hauser, Ben Parkes and Alan Gadian show that this initiative could have a positive effect on coral reef bleaching. The marine clouds would reflect sunlight more effectively and thus lower ocean surface temperatures. This would prevent coral death caused by rising temperatures.

Materials with a high albedo are already used to optimise the reflection of solar radiation. Swiss citizens have undertaken to cover nine of their glaciers, including the Rhone glacier, with large, white blankets to prevent them from melting. According to glaciologist David Volken, this could reduce melting by up to 70%. Research groups from the Heinrich Böll Foundation and the ETC Group support this initiative, opening new possibilities for the use of high-albedo materials, especially in construction.

Scientists at the Divecha Centre for Climate Change in India are seriously exploring the idea of launching an orbiting satellite that would reflect solar radiation. Even a 2% drop in solar radiation could make a big difference and considerably limit a rise in temperature. This project is therefore a great opportunity, but it requires the use of more advanced technologies than those we have today.

Absorbing CO_2 emissions through natural elements – such as phytoplankton or trees – has been well-trodden ground since the end of the 20th century. Just look at the buzz around reforestation projects in 2021. These include initiatives such as The Great Green Wall, which aims to create a climate barrier in the arid region of the African Sahel, thus halting the desertification process. The project should be completed by 2030 and may be a source of inspiration for similar initiatives in other parts of the world. The possibility of fertilising the oceans is also being explored as it could absorb between 30 and 50% of our CO_2 emissions.

What if there was a simple hack anyone could use to dabble with climate change? A whitepaper from Harvard Kennedy School's Belfer Center posed an interesting scenario for geoengineering taken out of government's hands. According to the piece, hobbyist kits for unmanned, high-altitude balloons can already be purchased for as little as $25. Imagine what would happen if someone launched a campaign via social media, calling on every citizen to launch high-altitude balloons into the sky. Each balloon would carry a small payload of particles that could reflect heat back into space – something that would be completely doable by mixing helium with sulphur dioxide.

Who knows, perhaps rogue acts like these – mild forms of green terrorism – will be exactly what our society needs to wake up and smell the coffee... brewed with a melted glacier? Think of it as democratising technology for the greater good.

PROBABILITY

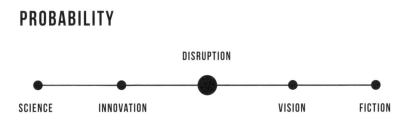

DISRUPTION

SCIENCE   INNOVATION   VISION   FICTION

Geoengineering encompasses a wide range of techniques to control the Earth's thermostat. Some, such as covering glaciers with large white blankets, are in use already and could easily be deployed on a global scale. Others are well under way but still require further research. Injecting aerosols into the stratosphere is theoretically possible with current technologies, but the right aerosol formula has yet to be found. The use of certain particles – such as sulphur dioxide – could actually be more devastating than beneficial.

Will we be using geoengineering by 2051 to try to counter the effects of climate change? Certainly. Getting a reflective satellite up and running may take some time and injecting aerosols into the stratosphere still need a few more years of research to come to life. These interventions also raise some interesting questions about who should be in control of how we tinker with the Earth's climate. Some even speculate that we could be heading for climate wars.

Either way, these methods give hope to humanity. It's comforting to know that there are ways to artificially halt the rise in temperatures while humans adapt to a new normal. It's exciting to know that not only could we maintain the current status quo of our beautiful planet, but even turn back the clock.

ALTERNATE REALITY LIVING

It's 6 pm, July 2051. A woman is standing in her bedroom, performing a strange ritual of hand movements. Somewhere between traffic cop and a conductor, she occasionally stops and taps her chest, then sweeps her hands into the air again for a final climax. This is Adeola, a commercials body music composer from Nairobi. She's wearing titanium monocle smart glasses – she likes to keep one eye on the real world. Using full-body motion-capture technology, she's composing a sound that's somewhere between a clarinet and a Sakara drum, using her fingers and her body as the instrument.

She's finishing work for the day, doing a final mix on a new ad for robotic chef's knives. Adeola mostly works in the B2R2C (Business to Robot to Consumer) space as that's where the decision-making – and the money – lies these days. Recent market research suggests that AI assistants respond well to rhythmic and melodic brand messaging.

She finally stops her day's work and walks over to her sparsely filled wardrobe. Everything inside is pure white. She pulls the kaftan she's wearing over her head and puts on a tight-fitting jumpsuit. She blinks twice to flick her monocle view to mirror view. What she sees now is a psychedelic two-piece suit made from recycled metaworld plastics. Yes, in augmented reality upcycling is still a thing.

Happy with her real-life AR look, she switches into the real world. She's hosting a small family gathering on her patio downstairs for her mom's 80th birthday. At her mom's request, they will enjoy 3D-printed burgers, pepperoni pizzas and strawberry milkshakes in line with a nostalgic 1980s theme. But the big surprise, the real bash is happening virtually. Friends from around the world will tune into an exclusive party venue in the Earth 2 megacity that now sits over London.

After months on a waiting list, Adeola got a last-minute cancellation for a space that resembles Wembley Arena. Here, she will surprise her mom with a re-enactment of the 1985 Live Aid concert. They will be standing shoulder to shoulder in a vast audience of holograms and – the clincher – meet avatar Freddie Mercury and David Bowie backstage.

THE UNREAL WILL BE OUR NEW REALITY

In 2051, the metaverse will be part of the minutiae of our daily lives. More than a quarter of the human population will derive income from these universes since the real economy has become heavily robotised. The massification of the virtual economy means that the production of real goods is dwindling, while trading in their virtual counterparts is skyrocketing. It's a world where it's much more difficult to distinguish between the haves and the have-nots. Someone might be living in a flooded slum in Ho Chi Minh City, but be a wealthy crypto trader. Or they might live in a real-life palace in Geneva and see their fortunes wiped out by factory closings and natural disasters.

What is the metaverse? An immersive, three-dimensional universe made up of thousands of virtual and augmented reality galaxies. Bigger than the web, it's the 'worldisation' of the internet, where physical and digital worlds have converged for good. Right now, we experience the internet when we go to it. Imagine if it's knitted into our daily lives, as inconspicuous as a squirrel in a tree. Fun, beautiful, entertaining, informative, useful – everything a great user experience should be.

Meandering between virtual and augmented reality will be as simple as switching a light on and off. By slipping on a pair of glasses or popping on some contact lenses, users can completely escape into a magical, shared virtual reality (VR) world. Here, your avatar can 'do anything and be anyone, without going anywhere at all', as *Ready Player One* so aptly puts it. In the virtual world, life moves fast and changes constantly, sweeping you up with the excitement of possibility as you bend reality in a game-like world. Your digital twin can have its own life or mirror yours. It can trial many things on your behalf, while it communicates with the IoT and wearables in your life. Want to see how you'll look after a two-week diet of grapes and herbal tea? Put your avatar on that regime and get realistic feedback.

Or you'll go through life in augmented reality, where every single thing in your peripheral vision comes alive with meaning. It's the real world as you see it – only better. Every object will have its digital mirror, whether it's a skin or interface. Everything is personalised and contextualised for ease of use – proactive rather than reactive. Even your oven might ask this: 'I see you have a cheese sandwich in your hand, can I switch on the grill for you?'

Some will leave the real world behind for good and live out their lives in a fantasy world of their choosing. Run an exclusive Earth 2 treehouse hotel deep in the former Namib Desert? Sure. Act out some politically incorrect fantasies in a *Westworld* type setting? Probably. Life as a cross-dressing mermaid? Whatever floats your virtual boat.

Others may well shun this cacophony of noise and visual overload and only immerse themselves in this world sparingly. You can't blame them, if you take Keiichi Matsuda's short film *Hyper-Reality* as an example. This is a world in which business and technology dominate – a bit like trying to find anything on an ad-peppered news site. Of course, the more you pay, the better the user experience will be. Users will choose from freemium packages, or wait to be invited to exclusive clubs. You will choose to live inside Apple's super-minimalist interface, Adobe or Snapchat's quirky and vibrant art-filled world or Google's street map-esque environment – filled with useful user-generated content.

TO EACH THEIR OWN (WORLD)

Could all this cause headaches? Definitely. Can't live without it? One thousand percent. A permanent meta-environment is coming and it's going to be huge. It goes by many names – the AR cloud, Azure, the mirror world, the Magicverse, the Spatial Internet, or Live Maps. It's going to be the new normal for how we work, play and socialise. 'Shall we eat in or eat out?' will become 'Shall we dine in real life or meet in *Animal Crossing*?' Your daily dog walk will become a treasure hunt through an enchanted forest. (A VR headset for your Labrador? Why not?)

Firstly, let's take the wearables. VR headsets have long graduated from gaming to enterprise applications (even warfare). The lineage of Microsoft's HoloLens can be traced back to an add-on for an Xbox game console. Today, these mixed reality smart glasses are used for everything from manufacturing and training to remote conferencing and surgeries. And let's not forget Google's Glass, which gives workers a huge boost in productivity by going hands-free.

Those are the glasses you wear at work – but what about the glasses you wear all the time? The ones where people can still see you wink and roll your eyes? Take your pick. The prescription specs Focals from Canadian start-up North, acquired by Google, look remarkably like the real deal. Facebook bought out microLED specialist Plessey to launch its version of everyday smart glasses. Designed by Ray-Ban manufacturer Luxottica, these are bound to look fabulous.

Meanwhile, Nreal has already struck a deal with Vodafone, giving European customers their first taste of a mixed reality future – right now. Users can shop, watch sport, enjoy AAA gaming experiences and more in augmented reality on a large 'virtual' screen in front of them. But the prize for the best-looking glasses must go to Snapchat. The company collaborated with none other than Gucci for its Spectacles 3 release. Featuring two cameras, it lets you create 3D content and embed AR lenses into your environment – as if they were always there.

With the wearables in place, the next bridge to overcome in completing the metaverse experience is mapping the world around us. Earth 2 is a virtual 1:1 scale version of our planet, in which real-world geolocations on a sectioned map correspond to user-generated digital virtual environments. These tiles can be owned, bought and sold. Very soon, these properties will also be customised in as much detail as you do your real-world environment.

Back in the real world, Big Tech is hard at work to bring users persistent shared AR experiences. Google and Apple's location anchors act as 'save buttons' for AR, enabling users to store the locations of their creations indefinitely. Microsoft does something similar in Azure Spatial Anchors, while Snapchat's Landmark AR uses location plus visual positioning to anchor AR content to buildings and monuments. The Los Angeles County Museum of Art was one of the first to collaborate with Snapchat, using augmented reality to create virtual monuments that explore the histories of LA communities.

PROBABILITY

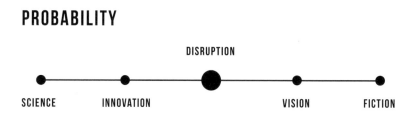

DISRUPTION

SCIENCE ——— INNOVATION ——————— VISION ——— FICTION

Still think the metaverse is only for gamers? A bit gimmicky? Every day, our world is edging closer to permanently living in an alternate reality. Take South Korea. It's one of the first countries to prepare a blueprint for a digital world. The nation's ICT ministry has launched an industry alliance to bolster the development of metaverse technology and ecosystems, with partners like SK Telecom and Hyundai. Samsung Asset Management, the country's biggest asset manager, has launched a fund tied to the metaverse.

Some people have wondered what on earth a car manufacturer is doing in the metaverse. If it's anything like BMW, it's already creating a digital twin of its factories and warehouses to use for product development. BMW harnesses Nvidia's metaverse platform Omniverse to create a digital representation of a physical asset, system or process. This way, the carmaker can virtually assess changes to its production lines before investing in the real deal.

We can only hope that, like Blockchain, the metaverse will be a true techno-logical democracy, with equality between all human and artificial beings. It's already providing a much-needed source of income in some of the poorest countries on earth. Take Venezuela, where the paper currency is worthless, but people survive by gold farming in *RuneScape*, exchanging coins for Bit-coin. In the Philippines, the play-to-earn model has been a lifeline for many who were hit hard by the pandemic. *Axie Infinity* players can earn yield in the form of tokens or other rewards, with many players earning three times the minimum wage. Such is the popularity of the earning model that there are now scholarships where Axie owners rent their NFTs to new players to learn the game and earn SLP (the in-game reward token) without having to invest any money upfront.

The bottom line – these fascinating case studies are already showing the world what can be achieved in a virtual world that isn't tied to currencies and real-world assets. The second world will make its own rules – where anyone can be a winner.

GENES
FOR YOUR JEANS

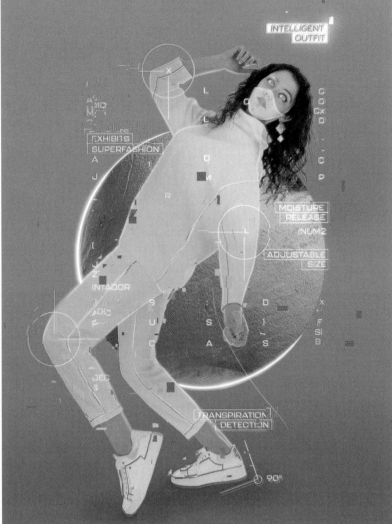

Today was the first time Bao ever touched jeans. What surprised him the most was not the roughness of the fabric, or the complete lack of fabric patches designed for specific body areas. It was the rigidity of the shape. He wonders how Grandpa ever felt comfortable in them. As he continues to clean out the attic – his grandfather having passed away from skin cancer earlier that month – he keeps sorting through objects he has never seen before in his life. They are mostly made from materials that seem weirdly crude to the twenty-year-old. Animal skins. Organic materials that required a lot of pesticides to grow, and from species that are now extinct. Bao organises everything in neat piles, ready for pick-up by the recycling plant.

Everything, except for the jeans. They intrigue him. On a sudden whim, he decides to try them on. Bao takes off his slim-fit trousers made from lab-grown fibres and puts on the old pair from his grandfather. He wriggles to pull them all the way up and it takes him a couple of seconds to remember how to operate the zipper. He's only seen one of those in a video where kids born after 2035 were presented with stuff from the beginning of the century and asked to figure out what the item was for. Zippers and buttons had become obsolete by the time he was old enough to develop a sense of fashion, around the same time spray-on clothing became mainstream, so he never actually operated one. When it all works out, he feels a small sense of accomplishment.

He takes his first steps around the dusty attic in his grandfather's trousers; the fabric feels odd and even chafes his skin a little. He wonders how badly this must have affected the skin of someone wearing these jeans for a whole day. The fabric clearly didn't adapt to the shape of his legs, which Bao considers a basic requirement for any clothing item. It should fit like a glove and gently envelop his body shape automatically. How were the probiotics in the fibres of the clothing item supposed to do their work if they weren't touching the skin? Then something else hits him. The smell! Musty, old and even a little mouldy. Clearly these pants did not have any odour-regulating features. They probably required regular washing, wasting precious water and contaminating water streams with harsh detergents. Bao struggles to take the trousers off and then drops them onto one of the piles for the recycling plant. He puts his own pair back on and is comforted, feeling the soft tissue on his knees. Shape memory instantly brings the fabric back to a perfect fit. He knows that the sensors in the fibres will quickly pick up on the mild chafing on his knees and activate the nanoparticles woven into the knee patches. The balance of his slightly sore skin will be restored, and the irritation will soon pass.

Bao puts everything in bags with a QR code on them and takes the bags out to the street where they will be scanned and picked up by autonomous vehicles that will take them to the plant. When he steps out of the house, his smartwatch beeps. It's an alert from the weather towers letting him know there has been a rise in UVA radiation since he entered the house at dawn. Bao quickly puts on his face-glove. Not only does this perfectly modelled piece of face-wear protect him from the sun, it also monitors the pH balance of his skin throughout the day. On his way home, he wonders why his grandfather ever settled for clothing that did absolutely nothing for his health or comfort.

YOUR CLOTHES SHOULD DO MORE THAN MAKE YOU LOOK GOOD

After the 2021 virus pandemic, demand for clothes to be functional, comfortable, and enjoyable to wear increased. The ever-so-conscious user became more demanding with every new year, yet brands didn't really take notice. That is, until their design departments reached a creative block; pressure from users forced them to turn to science, and to the invention of new materials and material blends.

The pressure was additionally amped up by the environmentalists, who deemed traditional organic materials like cotton or wool 'no longer sustainable', citing water shortage and the need to grow food on lands used for fibre production, while at the same time scrutinising synthetic materials like polyester and rayon. This ushered in a radical change in fashion, a shift from superficiality to sustainability and utility. In parallel, the cases of melanoma cancer rose, and skincare took priority over the frivolity of make-up. UVA and UVB ray fright resulted in the constant monitoring of radiation via smart watches and the inclusion of UV fighting ingredients in every single skincare item.

Fast-forward to 2051, where the brief idea of merging clothing and skincare is now a reality, and it combines nanotechnology and probiotics. Nanoparticles are being used to add hydrophobic, self-cleaning, anti-bacterial, UV-shielding and wound-healing properties to the fabric; probiotics are there to bring relief and care to the skin. Most applied nanoparticles became those of a mixture of silica coated with copper or silver due to durability, antibacterial and anti-odour properties, and zinc oxide nanoparticle blends for flame retardancy and UV protection.

To make sure these probiotics were properly absorbed by our skin, clothing needed to be skin-tight. That's when spray-on fashion became a thing. However, this type of clothing suited only those with little to no fluctuations in body weight (and body fat), which is why the sales of clothing with shape memory technology overtook that of the spray-on kinds in the late 2040s.

This new type of adaptable fashion was called 'bodywear'. It became very popular with mothers at first, since it didn't have to be re-applied like skincare and could be used for years. Because of its unique functions, an item of bodywear didn't have to be washed as often, an argument that wasn't lost on housewives with a limited budget. Nanoparticles and probiotics eventually became intertwined with new types of fabric, most notably spider silk. Because of its strength, the bodywear garments containing spider silk started being referred to as 'bodyshields'. It was not long after that, that the same multifunctional garments for the face, aptly named 'face-gloves', were widely available in clothing stores as well.

More advanced pieces of bodywear were soon separated into several regions, with specific skin demands addressed for each region. For example, the patches we were used to seeing on the elbows of men's shirts since the last century, were designed as of the 2040s to bring extra nourishment and moisture to that area, as most people's elbows are very dry. These patches were activated by sensors hidden within the patches, all the while communicating with the users' smartphones to deliver them live skin stats.

COME ON, WILL WE EVER REALLY USE CLOTHING FOR OUR HEALTH?

Though perhaps surprising, the prospect of using clothing as skincare is already being developed. In 2021. Fashion designer and scientist Rosie Broadhead announced she was partnering up with microbiologist Dr. Chris Callewaert (a.k.a. Dr. Armpit) to explore the possibilities of encapsulating probiotic bacteria into the fibres of clothing.

We're used to thinking of pineapple or fungus-derived materials as very advanced when the real material advancements stem from nanotechnology already used in the creation of nanomaterials. Examples of these materials used in this article have been researched for years, even though fashion has not caught up with them yet. An article from 2010 by Gerber et al. evaluated the effect of silver tricalcium phosphate nanoparticles onto Polyamide 6 fibres to build a reactive system against bacteria such as *Escherichia Coli* and *Streptococcus Sanguinis*, showing a killing efficiency of 99.99 or 100 percent within 24-hour contact time. In 2013, El-Hady et al. proposed a new, flame retardant approach based on the use of zinc oxide (ZnO) nanoparticles for their application to cellulosic fabrics (cotton polyester blend).

There are also selected wound dressing compounds being explored for the creation of wound healing materials. Among the overall compounds of great importance, Silver Nanoparticles (Ag NPs) are used as antimicrobials in hydrocolloids, alginates, and hydrofibrous materials fabricated by electro-spinning. Other relevant works focused on the preparation of wound dressing mats by electrospinning were presented by Hong and Rujitanaroj et al. in the late 2000s.

Fabrics we use today are far from sustainable – unlike spider silk, known for its strength and durability. Artificial spider silk has been successfully created in labs since 2009. The production consists of the extraction of the spider DNA that creates the silk and its transformation into microorganisms that eventually form yeast. As this yeast is being brewed, it creates a silk protein. These bio-fabricated raw materials remove the need for animals or insects and are far more efficient.

Futuristic clothing pieces made for the face are reminiscent of the popular Korean sheet masks that bring instant hydration or relief to the skin. An upgrade from those masks are machines with which users can already create their own DIY sheet face masks. The only problem with both of those masks is that they must eventually be removed. Ideally, in the future, we'll have face-wear or body-wear that will be hard to detect, providing the slow release of active ingredients throughout the day and not just upon – or shortly after – application. The detectability could be additionally improved or minimised by allowing users to have their faces scanned to create 3D moulded face-wear, much like Amorepacific already does for its users at their point-of-sales.

In 2013, Fabrican Ltd introduced spray-on clothing by presenting several clothing pieces that started out as a humble liquid. There have not been recent updates on the invention, but it's not hard to imagine something like this being implemented into the future of clothing as skincare. Especially since we already have spray-on makeup.

A company called Myant is pioneering the creation of clothing that can monitor our every move. Some people refer to what they do as producing 'smart fabric', though they prefer the term 'textile computing'. Yarns are paired with electronic sensors so that essential data can be captured from the human body. Another smart textile company creating similar clothing pieces is called Hexoskin.

The biggest challenge might be to create clothing that can shift, move and change temperature. Having a clothing piece that warms you when it's cold outside, keeps you snug when you're feeling in need of a hug, or changes patterns, sounds like a dream. Turns out designer Behnaz Farahi might be onto creating something similar. She combined 3D printing, shape memory alloy and computer vision technologies to make a top that moves, thanks to eye-tracking controlled movement. The top was inspired by involuntary body movements like pupil dilation, yawning or chills.

Changing the temperature should also become much easier. The brand Polar Seal creates tops that heat the lower and upper back on command, at the push of a button. Now, wouldn't we all love a hoodie that could get warmer when it's cold outside or snuggly and tight when we're tense, and that changes patterns based on our mood?

PROBABILITY

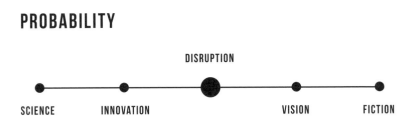

We have witnessed amazing advances being made in the field of fashion and nanotechnology, but their combined development and implementation is rather slow. Not only is it slow, but any attempts are quickly dismissed, as there seems to be a failure in seeing the added value in such innovations. Take spray-on fabric, for example. Ten years passed after the initial hype, with no significant updates made to date. Additionally, some papers on nanotechnology and nanoparticles are almost twenty years old, yet we haven't seen much representation of it in the field of fashion.

The issue with the spray-on fabric was apparent from the videos where the product in action was shown, often showcasing its benefits, but not how it would be removed. After removal, the person would have to have the skills to attach the two pieces back together, which was never demonstrated. On top of that, the clothing seemed fragile and not suitable for people whose weight might fluctuate, such as pregnant women. A great way to perhaps restart the conversation around spray-on fabric would be the introduction of spray-on, stretchy material that has the functionality and breathability of standard fabric.

Nowadays, we have amazing new organic fibres to work with, like pineapple and fungus, and soon spider silk will be within our reach, but the probability of those or more commonly used fibres such as cotton being intertwined with probiotics and nanotechnology doesn't seem likely, as most probiotics have an average shelf life of one year – if they're not in contact with air or water.

Nevertheless, it would be amazing to one day be able to wear a garment for a year without having to wash it. For now, we'll have to settle for probiotic-infused skincare. Longevity of the fabric is the current problem of 3D printed clothing as well. 3D printing materials are far more rigid than those used in the making of traditional fabrics, which makes them uncomfortable and easy to break. Those that aren't as rigid are still made of plastic – which really isn't helping the scrutiny fashion is under regarding sustainability. Ultimately, 3D printed clothing has yet to reach anywhere near the standards of comfort users demand from their garments.

Face garments or 'face-gloves' are a bit more promising, as Chinese bathers can be seen readily wearing a perfect beach combination of facial sunscreen and a hat called a 'facekini' when visiting beaches. Making it barely detectable and mouldable to a wearer's face, is the next step to making it closer to what Bao is wearing.

Although these ideas are not likely to wait for us in 2051, they're to be closely checked upon and followed. Changes might be coming slowly, but they are usually radical when they eventually do happen, at least on a fashion scale. After all, who would've thought that we would eventually be printing garments, after the invention of the first printer in the 1800s? All that is left to do to speed up the process of seeing these futuristic garments on shelves is for fashion houses to incorporate and hire scientists and microbiologists – with as much care as they do when hiring their creative directors.

CYBER
SMALL DEATH

Today will be her first time. Valentine looks at the thing on the bed in front of her with a little distrust. Will it feel as good as Rick made it sound? He seemed absolutely convinced she would love it, that it would make her body vibrate with pleasure and make her release with an intensity she had never experienced before. But the thought of surrendering her body so completely to the mercy of a machine controlled by someone she has never met in real life, made her second-guess the whole endeavour.

She tries to remember that she is here, today, in her bedroom with all that equipment, because she has grown to love Rick. Even though it was DNA mapping and clever algorithms that brought them together on the dating platform, her feelings for him are real. She remembers putting the thin stick in her mouth to get a cheek swab, such an innocent gesture, but the beginning of a big adventure. She thanks her lucky stars her biology and psyche profile proved a perfect match for Rick – she could have been matched with someone a lot worse... or no match at all in the database.

Valentine strips off her clothes and sits on the side of the bed. Her heart rate rises as she switches on the monitor that will connect the equipment to Rick's controllers on the other side of the ocean. Before she puts on the haptic suit, she wants to tease him with her real-life underwear. It's a black, see-through lace number she bought just for the occasion. When she sees his face on the screen, her heart races just like it did the first time she saw him in a practice session. He is just as handsome as his deepfake avatar she practiced easy conversation and appropriate flirting with on the dating platform. She had enjoyed it, even though it was mandatory before they could take their robots out on a string of amazing dates in the real world.

Rick reassures her, gives her a compliment, remotely tilts the camera so he can have a good look at her lingerie. Valentine notices his cheeks going red, which makes her bolder. She slowly hooks one finger behind a strap, drops it, then runs her finger down to her breast. She can hear Rick's breath speed up. Slowly, she tilts the camera down, lets it follow her hand further down until she slips it, teasingly, inside her panties. Judging by a quiet groan from the speakers, it has had the desired effect. 'Good', Valentine thinks, 'if we're going to use all this tech, I want him to know that I am as real as it gets, not some government-owned sexbot. She wants his thoughts to be only of her, the real Valentine.

In a low voice, Rick tells her how sexy she is, that he can't wait to please her. They both put on their simsuits and headsets at the same time. For a few seconds, they need to readjust to seeing each other's full body avatar up close. Then Rick smiles his disarming smile and reaches out his hand to touch her shoulder. Valentine feels the warmth of his palm, the pressure of his fingers – it feels quite real. Then his hand starts to descend. 'Take the Lovehandle', he breathes in her ear. Valentine feels around on the bed to grab the odd-shaped device. She pulls it closer as his hand reaches her nipple and squeezes it softly. A little gasp escapes her lips as the suit matches his grip on her body. 'Now hold it between your legs.' Valentine positions the device. 'I'm ready', she whispers. The handle begins to vibrate, slowly at first, but Rick controls the speed perfectly, based on the biofeedback he gets from her suit. As her pleasure rises, she clenches the Lovehandle between her thighs.' 'Come for me, oh yes, Valentine, come for me', he moans. And eventually she does, in exploding waves of intense pleasure.

21ˢᵀ CENTURY DATING

Like the millennia before, humanity's basic urges did not change. Humans still wanted to feed themselves, shelter from the elements, feel connected to others, procreate, procreate. The only thing that changed drastically over the years, was how technological inventions changed the way these needs were met. For example, humans started to farm food, then genetically manipulated it, then grew it from scratch in a lab.

Technology did the same for dating. First, people met in public places like at a dance, then they started meeting in online chatrooms, then services sprang up that claimed to match people with their soulmates based off a picture and a short bio. But by the 2030s, users of these types of services grew wary of their low success rate. Sure, a one-night stand was easy to come by – but what about a real connection, a perfect fit – both biologically and psychologically?

When the process of DNA profiling became quick and affordable to the wider public, dating services saw their chance. They started collaborating with personal genomics and biotechnology companies. Importing DNA data into a dating app was optional, but quickly welcomed by everyone looking for their perfect date. As it was possible to come up with multiple biological matches, an additional process was added to complement the search for the perfect mate through deep learning and mining the massive amount of behavioural data collected on dating sites in the last 30 years. In short: after a series of attempts to design the best way to match people, our society settled for DNA matching and AI personality models: self-developed algorithms superior over similarities, opposites, astrological signs or Myers-Briggs acronyms, to perfectly match couples based on characteristics they're compatible with.

When we were collectively used to more successful matches in digital dating, the majority of encounters still resulted in face-to-face meetups. But as technology developed further, this was no longer necessarily the case. Some relationships started – and stayed – exclusively digital. Those, of course, implied one of the partners being fully imaginary and virtually available 24/7. Deepfake technology came in handy for this one.

Dating simulation VR games became increasingly popular as well. Based on a user's profile, AI would generate partner profiles a user would be compatible with the most. That user could then virtually date in VR for a period. This became part of the offer – the more exclusive dating platforms offered deepfake counterparts of their actual members, so they could test potential matches hassle-free. The platform owners argued that this kind of simulation dating would allow users to know the likely outcome of relationships in advance. Or to train those who felt awkward and inexperienced in romantic situations so they were prepared once they met their real match. Or, in case of really high-profile users, to train potential matches how to behave in specific restaurant situations, how to make polite conversation, and how to move in more delicate social situations. In short, to have a better understanding of what to take note of when meeting their future, in-flesh partners and how to make sure the match would start off on the right foot.

Another type of dating that emerged, was hybrid dating, marketed towards introverts, people seeking privacy and relationships beyond the influence of the physical vessel or status and wealth. Hybrid dating became common in all dating apps and consisted of robots piloted by a real person going on a date with another real person in the real world. It allowed for a fair amount of privacy from the intruding lenses of the paparazzi. These types of *datingbots* would become popular amongst high-profile and political figures.

It wasn't until artificial intelligence was infused into those *datingbots*, that they started to have a wider societal impact. These bots were recommended to people suffering from chronic loneliness. Yet dating and companionship weren't the only things that changed. Intercourse did too. The porn industry has always been quick to innovate, and after the 2020s their inventions continued to trickle down into people's sex lives. 30 years ago, for example, no one would've guessed that sex workers would utilise Snapchat and Only-Fans, initially founded to be a membership platform like Patreon, to sell their services. In the 2040s, the porn industry was the first to use AI robots for personalised intercourse. Those wanting an advanced experience could incorporate a VR sex suit (referred to as *simsuit* in Valentine's story) to heighten the immersive aspect of their experience.

These AI powered robots could 'remember' data from previous sexual encounters (including from the *simsuit*, if one was worn) and their algorithms could use that data to calculate the best user experience for next time. But as these robots could also interact with their users in a way that simulated relationships, there was a sudden spike in human-robot marriages. The governments of countries where people could afford this expensive equipment, saw a radical drop in the number of births and had to step in. They created legislation that forbade the use of socio-emotional algorithms for sexbots. Other countries, where overpopulation or human trafficking was a big concern, governments legalised the use of AI powered sexbots – but only as sex workers. They issued standardised models, made them available in specific locations for standardised usage fees. It was not a huge success. The idea of the government having access to intimate data that could be traced back to an individual person, put a lot of people off.

Another measure that lots of governments took to decrease the emotional attachment to sexbots, was to forbid manufacturers to give their robots a gender identity. In a post-heterosexual society, no one really cared about gender labels, but their usage was more than discouraged. Features such as chest size or bottom attachments were to be sold separately, as users were not able to filter through robot models via a female or male gender filter.

In the end, dating and intercourse simply became a spectrum that ranged from fully natural to fully digital and everything in between. Those who could afford a more expensive model, like Rick in the story, ordered changeable parts and *simsuits* that could be operated at a distance. They used technology to connect with their loved ones across time and space, to fulfil that most basic need: to feel as close as they can to the people they love.

FALLING FOR A DEEPFAKE

The idea of robots goes way back in history, with Leonardo da Vinci thought to be the first one to try building a humanoid one in 1495.

Fast-forward to the 2020's, where we're witnessing Nao robots playing soccer, Atlas doing backflips, and Sophia being a full-blown Saudi Arabian citizen whose rumoured 'technical glitch' at an event in 2016 made her say that she'd destroy humans. Now those, including the Tesla bot that's in the making, are robots you've probably heard of. But there are other interesting android inventions that are just as exciting.

OriHime-D is an avatar robot, developed by Ory Laboratory. Using this robot, people teleworking or those with disabilities can control it to do physical work such as customer service, or to carry things. This is where the idea behind robots being used as vessels by private people for dating comes from. 'Why should humans only have one body?' is a valid question proposed by Kentaro Yoshifuji, CEO of Ory Laboratory.

Realbotix is developing the first sex robot, Harmony, with AI that monitors users' moods, preferences, and behavioural patterns, has synched eyes and mouth movements, articulated neck, and other features, all of which add up to the illusion of intimacy. Similarly, sex robot Samantha, created by Arran Lee Wright, uses AI to respond to different scenarios and even tells jokes. A recent update to the robot allows her to say 'no' when she's not 'in the mood'.

Miim is a robot that can walk, dance, and talk and its moving capability can be compared to those of Atlas robots. It has even walked a fashion show in 2009. It's to be assumed that some relationships will be exclusively digital, and with virtual partners. An example of how it translates to the time we're living in today is RealDollX app. With the app, you can make a unique companion character with a custom voice, looks, personality, avatar, and fashion sense. Users can talk, flirt and work on their seduction skills with the companion they made. While we're unlikely, to see real public personas with famous deepfakes like Tom Cruise or President Putin collaborate with these apps to enable users to date characters looking and sounding like them, it's not hard to imagine Z-list celebrities living off of selling personalised videos on cameo to eventually hop on the opportunity.

Dating simulation VR games are not new either. LovePlus was a 2009-released male-centred dating simulation game that allowed the user to have three girlfriends with unique personalities. The user allocated boyfriend points to each with the ultimate outcome of the game being to choose one of them. 2019 version of Sakura Wars introduced the dating simulator section to the game as well.

Combining dating simulation with VR isn't so distant either. VR Reality Sex Suits are already being sold, but in fact, aren't revolutionary at all. They combine VR-type headset we're familiar with, a game simulating sex, Tenga masturbator, and a pair of fake breasts. Nevertheless, looking ahead, we'll see a combination of VR headsets and haptic bodysuits. TeslaSuit, which has nothing to do with Musk's Tesla Bot, delivers haptic feedback, allows motion capture, and measures biometrics.

Perhaps the biggest surprise lies in the realm of DNA dating. DNA Romance is an online dating site that forecasts romantic chemistry between people using DNA markers that play a role in human attraction. They additionally forecast personality compatibility using psychology and allow users to evaluate psychological attraction based on their matches' photos and biographical details.

PROBABILITY

Havas' study conducted in 2017 revealed that 27% of millennials would consider dating a robot, with men being three times more likely to do so than women. But interestingly, according to 'Lovers Stories' survey, 69%

of millenials own a sex toy, and nearly half say they will add it to their collection in 2020. And as of recently, there has been a huge rise in women owning sex toy businesses and being actual users of them. Statista predicts the sex toy market will grow by almost 10% between 2019 and 2026.

Being that humans are amongst very few species that have sex for pleasure, and hence have sex more often than other species, having sex with robots could prevent unwanted pregnancies and sexual disease transmission.

Though it's hard to predict the rate of growth or fall in birth rate, in case it does grow, and it does so rapidly, we might see governments impose robots as solutions to slow down the overpopulation of the planet.

It might also reduce mortality associated with loneliness. A 2010 analysis of 300,000 people in 148 studies found that loneliness is associated with a 50% increase in mortality from any cause. This makes it comparable to smoking 15 cigarettes a day, and more dangerous than obesity.

Having sex with robots might find some objections at first from the sex workers, until they realise it's a lot safer to do sex work online anyway, while the rest is indulging in sexual activities with robots or their partners. Users are already joining websites like OnlyFans and creating their own websites where they sell explicit content. It is likely we'll see more of them completely detach themselves from their sex worker personae by using deepfakes while providing voiceover in the content they make. Some of them were already doing so until OnlyFans announced they'll be banning sexually explicit content and deepfakes.

This, however, resulted in creators leaving their platform for others where they can still share the content they produce. Sex workers have been using Snapcash to earn money via Snapchat, Youtube to promote their websites, and Twitter to post nudes way before OnlyFans was in the game.

It's not hard to visualise the 'vision' since most of the tech is already here: the robots, the haptics and the VR. It's only a matter of how fast humans will adapt to this new, artificial and robotic sexual environment.

AFRICA, THE SUPERNATION

A dapper young gent is standing on a helipad at the top of his building, impatiently waiting for his date. It's a swelteringly hot day; the temperature hovering around the 45°C mark. Luckily, he's wearing a special smart jacket that syncs with the thermostat inside his apartment to keep him cool. The dusky sunset plays a stunning light show on the biomorphic, parametrically designed buildings around him. If he peers down, he sees city dwellers milling around in lush, pedestrianised spaces. He taps his smart glasses into view to see where his beloved could be. The interactive map view shows that she'll be there in five minutes. He sighs; they will be late as usual.

A few moments later, a whirring, jet-black Skyride swoops in and lands softly on the helipad. These flying cars are now all the rage, with a look somewhere between ride-on drone and air moped. Is this a deleted scene from Luc Besson's The Fifth Element? No, just another day in Africa. The year is 2051. The happy couple is Aadi and Loide who live in Walalapo, a sprawling smart city between Namibia's port towns of Walvis Bay and Swakopmund. Home to over four million inhabitants, it's one of Africa's latest tech and blue economy hubs. Paying homage to Akon City architect Hussein Bakri, the cityscape is a futuristic Wakanda-like vision of undulating high-rises and man-made lakes with prolific mangroves and seawater engineered plants.

The power couple is on their way to a class reunion in Kgale Lake City in Botswana. Loide is dressed to kill in a skintight, sun-reflecting leotard and Adidas 'living' moss trainers, while Aadi (true to his brand) is matching his smart jacket with a flowy kelp fabric robe in deep emerald. The couple loves to joke at dinner parties that they are both miners. Loide made her bitcoin millions as an early adopter crypto miner and now works as a much-in-demand metaverse stylist, while Aadi specialises in (eco-friendly) mining of the ocean. His company transformed the Namibian coastline with many marine biotech firsts. He was the first African to create a megafarm cultivating GM corn with seawater, and also the first to pioneer sea farms deep in the Atlantic Ocean.

Loide gets her road trip playlist ready. The couple will hop on a high-speed train to Botswana, without worrying about border controls. The long-desired *'visa-free Africa' finally started coming together in the 2040s, with at least half of Africa's nations supporting the common African passport. The five-hour journey is a great time to catch up with other friends and debate about the latest blockchain tech making waves. They love watching the landscape change and seeing the savannah woodlands near Mount Erongo, where they used to go game-watching in their youth.*

MEET THE IAFRICANS

By 2050, Africa will be well on its way to becoming an economic superpower. With 40% of the world's youth population living, working and spending their way to greatness, more than half of the continent will be under the age of 25. If Generation Alphas are called the iGeneration, the most tech-savvy young people the world has ever seen, can you imagine how forward-thinking the 'artificial' Generation Betas (born post-2025) will be? These two generations will be the wealthiest, most educated, and technologically literate demographics in history.

Of course, sceptics will be quick to say it won't be like that for someone born in a Khartoum slum or on a tiny farm in Angola. But why not? Mobile phones are ubiquitous in Africa. Mobiles are the reason the African youth is a leapfrog population. They skipped landlines and computers and went straight to running their lives on mobile phones. Right now, there are more mobile phone users in Africa than in the US or Europe (approx. 650 million). More people have access to mobiles than to clean water, a bank account or electricity. In countries like Nigeria, Kenya and South Africa, nine out of ten people own a phone (or several phones).

A mobile phone is a lifeline: bank, landline, email, social hangout, recruitor, entertainer, doctor, teacher, tutor. Africa is not mobile-first, it's mobile-only. Through their cell phones, young people seamlessly connect with their neighbours and to the global economy. Boasting some of the highest grassroots adoption rates for crypto in the world, Africa is ripe for a crypto revolution. These mobile-native iAfricans will lead the way.

Growing up on the 'children's continent', they'll wear health trackers from birth and use AI diagnoses and treatments to live healthier lives. E-learning will be part of life. Interacting with voice assistants will be completely natural. Kids who can afford it will play with their smart, internet-connected toys and spend most of their time wearing virtual reality headsets. Youth culture will drive Africa. Young entrepreneurs and creatives will continue to break new ground, from Nigeria's famous Nollywood to Kenya's Silicon Savannah. The middle class will grow to 1.1 billion by 2060 and account for 42% of the predicted population of four billion. This means Africans living below the poverty line will be in the minority at 33%, according to the African Development Bank.

Money will change hands effortlessly across borders, without the need for bank accounts or even fiat currencies. Following the examples of Senegal, Tunisia and Nigeria, countries will roll out central bank digital currencies (CBDCs). In just a few decades, domestic CBDCs will come together to form Africa's long-desired monetary union – designed by Africans, for Africans. It may be tied to a common blockchain, who knows? Africans may even decide to back their own cryptocurrency tied to its abundant food and green energy industries. Remember when Colonel Gaddafi wanted the African continent to switch to the gold dinar, diverting its oil revenues to funds controlled by the state rather than by American banks? The idea may seem extreme but when the power shifts, anything could be possible.

When the rest of the world goes down the path of isolationism, Africa's free trade area will become a global game-changer – a model of cross-border cooperation for shared prosperity. With 60% of the arable land left in the world, Africa will be the laboratory for testing new approaches to boosting food production. This will go hand-in-hand with a massive infusion of private money, technology and infrastructure. As investment floods to rural areas, local farmers will learn new skills and co-exist profitably with industrial farming. Through precision agriculture, Africa will not only learn to feed its own two or three billion people but will also supply many of the world's food-scarce countries. Get it right, and Africa will rewrite the narrative, going from economic backwater to resource-rich superpower. Get it right and it will be the success story of the 21st century.

1.2 BILLION–PERSON MARKET ON THE CUSP OF STRATOSPHERIC GROWTH

Home to seven of the world's fastest-growing economies, Africa is on the move. Consumer spend is already four trillion US dollars, having grown 5% a year for the last two decades. Africa has the people, the resources, and the sheer will to transform. How? Through eager adoption of digital currencies, a race to feed itself (and the world) and smart green energy, as well as through rapid urbanisation, rising incomes and a young, growing population. Soon, Africa will be the fastest-urbanising region in the world. With as many cities with over one million inhabitants as North America, more than 80% of its population growth over the next two decades will take place in cities. Currently, the per capita income of Africa's cities is more than double the continental average, making them most attractive markets to investors. McKinsey predicts that, by 2030, Africa will have 17 cities with more than five million inhabitants.

When it kicked off the African Continental Free Trade Area (AfCFTA) in 2021, Africa not only created the world's largest free-trade area since the founding of the World Trade Organization in 1995. It also laid the best ever foundation for economic growth, industrialisation and sustainable development in Africa. With 53 countries signed on, this single continental market for goods and services is well on its way to establishing a continental customs union. An investment, economic diversification, and job creation blueprint that will shape the future of Africa in the years to come, AfCFTA will alleviate poverty and improve economic gains for women.

There are many who feel that the AfCFTA can only fully reach its potential with a unified African currency. One money market could mean the free movement of people, capital, goods and services, accelerated industrialisation, and becoming better known for exports than imports. One giant leap towards fixing Africa's fragmentation was the formal rollout of the Pan-African Payment and Settlement System (PAPSS) in 2021. This has been spearheaded by the continent's trade finance institution, the Africa Export-Import Bank (Afreximbank). PAPSS will eventually enable instant payments – in African currencies – between merchants on the continent, which will help meet its ultimate goal: reducing Africa's dependency on currencies like the dollar and the euro.

But perhaps the eventual, unified currency will be crypto. After all, crypto is not bound by geography because it's internet-based, with all transactions stored in the blockchain. Developing economies lead the world in crypto adoption and Africa is no different. It shows great promise because of its unique combination of economic and demographic trends. In sub-Saharan Africa alone, as much bitcoin is sent via peer-to-peer exchanges per month than is transferred in North America (more than $80m). Chainalysis estimated that Africa's cryptocurrency market grew by 1200% between July 2020 and June 2021.

Africa's young, fast-growing, mobile-using population thinks nothing of bypassing banks and trading money in unique ways. All you need is a cell phone signal (not even Internet) to send and receive money. In countries without an expansive banking infrastructure, mobile money apps like M-Pesa and MoMo help Africans meet their basic needs. Mobile money is so huge, sub-Saharan Africa accounts for half of live mobile money services and two-thirds of total transactions in the world. In Kenya in 2019, almost half of the country's GDP was moved through mobile phones.

IT'S ALL ABOUT THE NEW MONEY

If people are already used to transacting with mobile wallets, crypto is the next logical step. Kenya's BitPesa was created with a vision to reduce the cost of money transfers between African currencies and reduce the reliance on legacy financial systems. Nigeria's SureRemit is another blockchain company that wants to turn the remittances industry on its head. Users purchase vouchers to send to friends and family – instead of fiat, taxable currency – which can be redeemed for real-world goods and services.

When El Salvador made history by becoming the first country to make Bitcoin legal tender, experts were quick to predict that African countries will soon follow suit. If you lived in a country with double-digit inflation like Zimbabwe, Ethiopia, Angola, Zambia and South Sudan, wouldn't you also want to be part of the global financial system? Artist Akon has the right idea. He founded his own cryptocurrency, Akoin, and together with the Senegalese government, is behind the futuristic six-billion dollar Akon City. Besides being spectacularly

Wakanda-esque, this smart city will be an interesting experiment to see whether digital currency can be better integrated into Africa as a whole. Everything from schools and universities to malls and property development will use the Akoin. Architect Hussein Bakri has created a magnificent vision for this multi-purpose city – complete with a parking lot for flying cars.

Akon believes that crypto is the foundation for amazing ideas to manifest. 'Crypto puts power back into the hands of the people by creating a transparent environment where trust can flourish. Decentralisation and cryptocurrency will allow for a new financial framework to develop on the continent, empowering entrepreneurs to build and execute their businesses digitally across borders.'

When Akon isn't planning sprawling cities – one for Uganda is already in the pipeline – he is also bringing low-cost, sustainable electricity to Africa. Since 2014, the Akon Lighting Africa project has provided 25 African nations with solar-powered electricity via street lamps and solar panels. The project has a big goal: to provide solar-powered electricity to 250 million people by 2030. So don't knock Akon City – Akon Lighting also seemed impossible at first.

It's not just bitcoin and Akoin trying to get a foothold in Africa. The Afro was founded in 2018 but has struggled to become Africa's bitcoin. South Africa's Safcoin has had better luck, making history as Africa's first proof-of-work coin; going global with a listing on HotBit. Safcoin has also launched MobiJobs – Africa's first blockchain-powered micro-jobs platform, connecting businesses and gig economy workers across Africa, and Cryptovalley – an e-commerce platform allowing small businesses to sell and pay for products with crypto.

African blockchain companies are also doing incredible things. There's Wala, bringing financial services to the unbanked. SunExchange is a marketplace that allows anyone in the world to invest in solar energy projects using bitcoin. Blockchain is also used for land registry services, such as Ghana's Bitland, and agriculture – like Agrikore, offering a virtual marketplace for buyers and sellers. Free from legacy baggage, African blockchain tech is starting from scratch, bringing unique opportunities for a pan-African economy.

AFRICA'S GREATEST ASSET? ITS PEOPLE

Their ingenuity and spirit, for sure, but also the sheer number of them. The size of Africa's working-age population is expected to surpass both India's and China's by 2034. By 2050, a quarter of the world's population will live in Africa. So, when Asia's working-age population declines, Africans – millions of them – will be ready to fill the gap. China will shift 100 million labour-intensive manufacturing jobs offshore by 2030. One can just hope that the lion's share of that comes to Africa. China has deepened its links to the continent, investing heavily in industrialisation, infrastructure and agricultural innovation. The world leader in solar energy technology, China helped upgrade solar capacity in Africa from 739 megawatts to 5,500 MW over a decade. Wind energy installations during the same period jumped from just 108 MW to 6,100 MW. Africa's abundance of sunshine will remain one of its top strengths. It has seven of the ten sunniest countries on earth: Chad, Egypt, Kenya, Madagascar, Niger, South Africa and Sudan. South Africa is already one of the world's top ten producers of solar power, and there are massive developments in Kenya, Rwanda, Ghana and Uganda. Morocco's Noor Power Plant is the world's largest concentrated solar power (CSP) plant that hopes to one day supply Europe with power. So easy was it for Morocco to meet its electricity needs that it has already exceeded its 35% renewable energy target and has now set itself a new target of 52% by 2030. Imagine one day charging your smart home with power from Morocco.

These colossal plants are expensive to develop and maintain, but don't think for a moment that solar power only comes in bulky, silicone panels. South Africa's Helio100 could become the smallest, most cost-effective plug-and-play solar solution in the world. Using CSP technology, it uses a field of tracking mirrors (heliostats) and a small tower to capture concentrated sunlight. Just 100 heliostats of 2.2 square metres can generate 150 kilowatts of power in total – enough for ten households. Tanzania-born Zola Electric is another company offering affordable ways for African homes and businesses to get off the grid. It's backed by the founders of SolarCity (now Tesla Energy).

THE SUN SHINES BRIGHTLY ON AFRICA

Perovskite solar cells are another much-hyped alternative to silicon. It promises to be less expensive and more efficient than silicon, and many companies are already taking their models to market. Very soon, these cells could reach the holy grail of 30% efficiency. Perovskites can be made to around 300 nanometres (much thinner than a human hair), which means a coating could one day be applied to anything from buildings and cars to clothing and cows.

Did we say cows? Saule Technologies in Warsaw has come up with an ink-jet printing process for manufacturing perovskite solar cells encased in a flexible plastic. One of its recent projects in Ukraine was to track two bison wearing telemetry collars powered by perovskite solar cells. The company also offers Internet-of-Things solar-powered price tags, solar blinds and e-mobility friendly car ports. It's safe to say that in just a few decades, everything that moves will be solar-powered. We may well all hang solar sheets from our washing lines or make our dogs wear solar jackets to power our morning cappuccino. If this tech can reach the world's sunniest continent, we'll light up the world.

As green technologies are scaled in Europe, nations will increasingly reach out to Africa for more cost-effective green fuel. Green hydrogen, for example, plays a big role in Germany's climate protection efforts. The government wants to partner with West Africa to jointly ramp up hydrogen production. It believes that with three-quarters of West Africa's land being suitable for wind turbines (and electricity production costs being half that of Germany), it could be a win-win collaboration. By using wind and solar energy, West Africa has the potential to generate up to 165,000 terawatt hours of green hydrogen per year – about 1,500 times Germany's estimated hydrogen demand for 2030.

WHAT IS PLAN B WITH THE WATER

Green energy and the infrastructure that comes with it, will create millions of jobs. But how will all these mouths be fed? And what about the water? At some point, someone will point out that while Africa has enviably fertile lands, it's also staring down the barrel of climate change. The weather is going to get hotter and hotter, and La Niña will continue to bring terrible droughts. The devastating droughts of southern Madagascar are just a taste of what will likely play out in Africa over the next decades. But even when rainfall dwindles, smart people find a way around it. Droughts in California have forced farmers to experiment with dry farming (no irrigation), while in Wyoming they've turned to planting the perennial wheatgrass Kernza. Deeply rooted, this cousin of wheat boasts improved yield, seed size and disease resistance – and it's lower in gluten to boot.

Then, of course, if there isn't fresh water, there's always seawater. California start-up Agrisea has made it its business to come up with salt-tolerant crops to fight hunger. Identifying the genes in salt-tolerant organisms like mangroves, they are now growing rice in ocean farms in Vietnam – one of the most notoriously resource-intensive grains in the world. Besides giving water-scarce countries a solid plan B, this method of ocean water farming is incredibly exciting. To meet the demand for rice alone in 2050, it's estimated that we'll need land equivalent to the size of Chile. If this method can be perfected, anyone from any country can become a rice tycoon. As Agrisea's CEO, Luke Young, predicts: 'In 30 years' time, it will be normal to see large floating crop islands off the coast of many countries, growing a variety of crops suitable for the marine environment.'

Our palates will change, and new food sources will appear on our plates. In Morocco, SuSeWi uses innovative technology to produce microalgae on land using only sun, sea and wind. It has the world's largest algae growth pond, a 30 000 square metre production facility. If you've been following food trends, you'll know that this green sludge is the future of food: rich in nutrients and able to revolutionise the world of nutrition, medicine and cosmetics. For now, our palates only know nori, wakame and spirulina, but in 2050, kelp burgers and algae pesto will be our staples.

The bottom line: if water is an issue, you must farm differently. You need to farm better. No drinking water? Let's get the Southern Ice Project back on the table. This ambitious plan was hatched by salvage expert Nicholas Sloane to tow an iceberg from Antarctica to Cape Town, to combat an impending Day Zero. Back in 2018, after three years of severe drought, the city was at risk of becoming one of the first in the world to run out of municipal water. The icebergs are melting anyway, why not? The plan may seem extreme but it's nothing new. In the 1800s, breweries in Chile towed small ones to use for refrigeration, while in the 1940s there were wild plans to transport an eight billion-ton iceberg to San Diego to assist with California droughts.

AFRICA: THE BREADBASKET OF THE WORLD?

It's hard to imagine now that a continent with so much hunger can one day supply other continents with food. But the enormous 70% increase in food production required will need to come from somewhere. Other continents simply won't have the agricultural resources to deal with this challenge. According to the UN's Food and Agriculture Organization (FAO), there will be a 21% increase in agricultural and fish production between 2020 and 2029 in sub-Saharan Africa. With the support of eager, local governments, the number of land deals has soared, and we've seen an unparalleled transfer of land ownership. Some will say this is simply a new form of neo-colonialism, but it certainly has the potential to increase food security, employment and income generation.

PROBABILITY

If Africa can move past its two biggest commodities – corruption and organised crime – it could have a bright future. The big difference with an African economic renaissance? Everyone's in on the action – every income group and, eventually, every country from Zambia to Mali. Africa is doing pretty well trading with itself, thank you very much. McKinsey reports that two out of every three dollars in new African economic activity come from goods and services sold within Africa itself.

While Africa's dream of a continental high-speed rail and road network is being realised, its tech ecosystem is powering trade like you can't imagine. Best of all, this growth was actually accelerated by the Covid pandemic. There are no rules here; no competitors to speak of. This is sky's-the-limit territory. It's even somewhat under the radar. Says the *Financial Times*: 'One sometimes-overlooked reason behind rapid growth is that there are virtually no legacy tech-enabled players already on the continent. In logistics, agriculture, medicine and B2B financing, there are no existing incumbents offering decent, first-generation technology platforms to compete with. Kobo "competes" with agents often reached unreliably by phone only. Twiga "competes" with stallholders having to make 4 am trips to wholesalers to buy the day's produce. And mPharma "competes" with an archaic distribution network prone to selling fake medicine at sky-high prices. The same reason these companies can scale so quickly is the same reason they are so important.'

With less than thirty years ahead to 2050, Africa is on the same growth path as China, Korea and India 20 or 30 years ago. But to think it will grow along the same trajectory would be a gross underestimation. Africa is ideally positioned to leapfrog centuries of industrial development and benefit from the achievements of the information age. In other words, expect nothing, expect everything. Watch this space.

PIMP
MY DNA

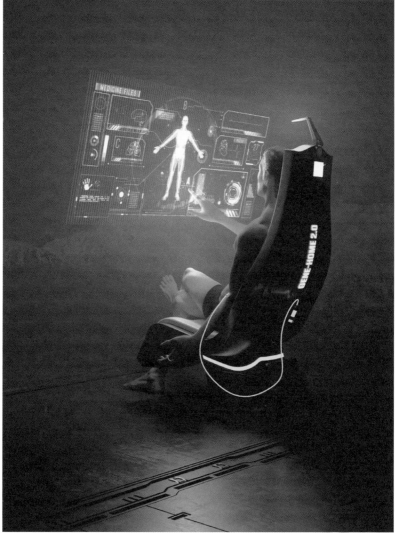

Stephen and Laila are the happy future parents of Lani, a little boy. The pregnancy is almost full-term, and the couple is heading to their last pre-natal appointment with their child's geneticist. Dr. Austin has been a true angel to them since the first appointment. At an early stage of the pregnancy, the foetus wasn't developing well as Laila was found to be carrying the gene for sickle-cell anaemia.

But now there is nothing to worry about. As soon as the problem was detected, the geneticist performed all the necessary procedures and the couple is now preparing for the birth of a healthy child. With both parents being huge endurance sports addicts, Dr. Austin has also worked on a few genes – namely ACTN3 and ACE – which will favour the future child's strength and endurance development. The future parents will, of course, let their child choose whatever he wants to do, but they believe improved endurance and strength will help him in many ways in life.

Fast-forward to summer, 2051 – Tahiti. Lani has grown up so much already. He's now a healthy five-year-old boy, enjoying his childhood on the island. On this Wednesday morning, he's meeting his friends for his swimming lesson at the local surf school. The little boy has shown great capacity. He already manages to keep his head above water and has mastered breaststroke. The genetic modifications his parents made before he was even born means that – compared to some of the other kids – he is almost tireless, and he tries over and over until he succeeds. Ironically, if it weren't for his development problem, Stephen and Laila would never have thought to consult with a geneticist. It's a trend that has only become popular in the last three years.

Once the lesson is over, Stephen takes his son home for lunch. Thanks to his genetic profile, established since his birth, Lani's parents always know what to give their child to help him achieve optimal growth. They know the exact amount of proteins, carbs, fats and nutrients he should take in each week, and as they have now registered their child's affinity for water sports in his genetic profile, they also supplement him with vitamins B and C. Nothing too serious for now, he's just a child after all.

This afternoon, Lani has his bi-annual check-up with the same geneticist his mom had consulted with during her pregnancy. There's no need to go to the lab, as Dr. Austin calls them in holo-conference. Thanks to constant updates from a chip implanted in Lani's shoulder, Dr. Austin has access to all data related to the child's genetic development. Everything seems normal. And if, as Lani grows up, he becomes more serious about water sports, thanks to the advance of genetics research, the doctor will even be able to improve his metabolism. She might also even be able to help him breathe underwater for up to two minutes. With such rapid progress being made in the field of genetics, there's no knowing what they might be able to do next.

FROM UNDERSTANDING TO ACTION

After spending many years studying genetic biology, scientists began to use what they learned in the 1970s to venture into the promising field of genetic engineering. The first genetically modified rat – the work of Beatrice Mintz & Rudolph Jaenisch in 1974 – paved the way for an extraordinary human adventure: DNA editing.

In 2021, with the completion of the 'human genome project', biogenetics takes another giant leap forward, triggering revolution in the medical field. At the same time, similar advances are being made in the fields of agriculture, through the sequencing of the DNA of the plants and animals that populate the planet. It has long been known that DNA is the key to understanding and changing the world. Now that we have a very precise idea of how it works, it is possible to take control of it to improve our daily lives. All of which means better treatment of diseases, better agricultural yields, as well as a grip on Mother Nature that will improve our living conditions and therefore our longevity.

With the exponential evolution of technology, the archaic and cumbersome equipment needed for gene editing is rapidly giving way to instruments that are increasingly sophisticated, but also more compact and transportable. There is no longer any need to build imposing laboratories to practice the discipline. It is now possible to do so from anywhere, and more affordably.

In parallel with this, genetics will also benefit from the giant steps made in the field of artificial intelligence. Gradually, the machine will take over and reduce human interaction to a strict minimum. Now capable of formulating hypotheses on its own, it can also design simulations, analyse the data, and draw the necessary scientific conclusions.

POCKET LABORATORIES

If, at the beginning of the 20th century, it was already possible to obtain DNA and RNA codes online, or to consult collaborative genomic databases (such as the NCBI or the UCSC), the decades that followed saw the emergence of a sort of 'collective genetic consciousness', facilitating the convergence of all individual knowledge in the field. From being localised, experiments have become global, with scientists from the four corners of the planet now able to collaborate remotely, in real time, on the same experiments.

Genetics used to be the preserve of large laboratories, but today it is practised in our kitchens. Now accessible to everyone, everywhere, it has experienced unprecedented growth in recent years, fuelled by better sharing of global knowledge, the easy exchange of ideas and designs, and the undeniable contribution of increasingly sophisticated artificial assistants and advisors. From the early days of CRISPR technology, to today's automated editors, genetics is undoubtedly one of the areas of science where progress has been most notable, and most beneficial to our civilisation.

We can now directly influence things like intelligence, stamina and life expectancy, to improve the living conditions of the entire human race. At the same time, we are preparing future generations for the challenges of tomorrow, by intervening in the genetic codes of our fauna and flora. With the phenomenon of designer babies now becoming widespread, it is now possible to not only reduce potential health risks at the embryonic stage, but also to accentuate certain physical characteristics of future new-borns, leading ultimately to the improvement of our entire civilisation.

THE ADVENT OF CHIMERAS

Of course, since its emergence, genetic engineering has had its detractors – those worried about the inevitable crossing of an indefinable ethical threshold. But the inevitable democratisation of the technology has not really left time for debate. The highly controversial experiments involving the combination of human DNA with animal DNA, the famous 'chimeras', have nevertheless led to unprecedented medical advances. The laboratory design of mice, with an immune system modelled on that of humans, has made it possible – through pharmacogenomics – to study the effect of certain drugs without the pitfalls of using humans as guinea pigs. And if the idea of improving our intrinsic characteristics by importing the specificities of certain animals (to increase our field of vision, or to improve our metabolism) is still frowned upon by a large part of the population, this is nonetheless the future of genetics. In the long term, these artificial evolutions will make it possible to facilitate the colonisation of other worlds, by 'reconfiguring' our respiratory system so that it is able, for example, to breathe the air of otherwise hostile exoplanets.

PROBABILITY

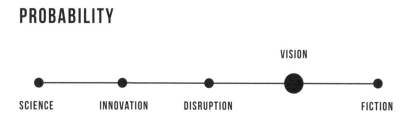

VISION

SCIENCE — INNOVATION — DISRUPTION — VISION — FICTION

Medicine, agriculture, the environment, the climate – the fields of application of genetic engineering are numerous, making it an indispensable discipline for the advancement of our civilisation. The main obstacle to its development will clearly not be technological. Most of the tools needed to master it already exist (virtualisation, data sharing, AI) or are on their way to becoming a reality. The main obstacle to its development will undoubtedly be the ethical and moral considerations – to which we do not yet have a definitive answer. A pessimistic vision of the misuse of genetic applications by ill-intentioned people is contrasted with a more humanistic and optimistic vision that sees an exponential increase in the number of advances beneficial to humanity. The environmental implications of DNA editing should not be minimised; most genetic experiments today are conducted by private companies not exactly known for their concern for the planet. And the risks of side effects are – for the time being – still rather difficult to measure (crossbreeding, destruction of ecological systems, etc.). Not to mention the potentially colossal transformative impact on the very essence of what makes us human.

To solve this ethical problem, we will need to combine these technological advances with other areas of reflection. It may require us to seek the combined wisdom of philosophers, economists, historians, lawyers and politicians to help solve this complex conundrum. To serve humanity in the best possible way, genetics must become and remain everyone's business, not just the preserve of technicians.

ENTER
THE ARENA

UPPER VIEW   PLAYER LOCK   BALL LOCK

July 8, 2051. What a sunny Saturday! The ideal weather for this year's Android Ultra Soccer World Cup final. They were less lucky at the Cyborg's edition, which happened just before. The augmented human teams had to deal with terrible weather conditions for almost every game. However, despite the great weather, Tom is staying at home to watch the game. But don't pity him! As the kick-off approaches, he sits comfortably on his sofa and pours himself a beer. He activates his neural implant with a furtive thought and in a nanosecond, he is propelled to the stands of the virtual Net-Wembley Stadium. The smell of the grass tickles his nostrils, and as he looks around, he sees the whole crowd of other aficionados, like him, connected remotely thanks to their neural implants – but scattered all over the world in real life. The illusion is perfect. Tom feels the way he did when he was physically going to the stadium as a kid, to watch the games of his favourite club, Manchester United.

After virtually reaching the stands, Tom silently orders his implant to show him the panoramic view of the pitch via one of the hundreds of nano-drone cameras recording the event. From this vantage point, Tom can check all the information scrolling, such as summaries of the previous games, the weather, the physical condition of the players and the predictions. Of course, there's also a whole lot of targeted advertising – betting platforms, food, tickets for upcoming games, you name it. Tom isn't letting himself become distracted by these. He nervously scans the AI predictions regarding the outcome of the match, then slides to the settings menu to select the voice of the virtual commentator who will cover the game.

One last look around the stadium, one last look at the stats, and he's ready for kick-off. The tension is palpable, especially as Tom is not just another spectator today. For a small fee, he has earned the right to participate in the game by remotely controlling one of the androids provided by FIFA. Tom puts down his beer and wipes his forehead. Kick-off is just five minutes away now.

As the androids enter the pitch, the crowd starts cheering for their favourite athletes. Today, the South Korean FC Jeonbuk Androids are playing against their Manchester United counterparts. Of course, Tom will do everything to ensure Manchester comes out on top and will play his role of remote defender very seriously. The two teams take the pitch; the serious business is about to begin. When the whistle blows, Tom automatically changes his point-of-view to that of his android's! The game is underway!

All too quickly, the half-time whistle is blown. Tom is having such a blast that he feels like the game started just a few minutes ago, even though the South Korean team is one goal ahead. The cheering crowd, the sensations on the pitch, the nano-drones transmitting the game all over the world – everything's like a dream. His android goes down to the stadium's changing rooms, where its mechanics will be checked by an AI to make sure it works until the end of the game. All the data from the first half is collected by an AI trainer. Valuable insights will then be given to the remote players, including, of course, all the weaknesses of the opposing team.

The game is over. Tom is a bit disappointed. Although his team lost, they definitely held their own against the defending champions. He de-activates his n eural implant and lies on his couch, exhausted but happy he could experience 90 minutes in the life of a high-level athlete. He will extend his sporting experience by attending the closing show happening on the metaverse to thank all the tournament's participants: a huge live tournament of Cyberwatch.

HOW MANY ANDROBOTS DOES IT TAKE TO SCORE A GOAL?

Until the beginning of the 21st century, the world of professional sports, known as high-level sports, had been reticent about the active incursion of technology into its various competitions. Already, with an increase in the collection of metrics to analyse athletes' performances *a posteriori* and improve their equipment, slow-motion video to study possible rule violations, or simply augmented reality to display advertisements or statistics, it was unconceivable at the time that technology could play a much more substantial role in the physical performance itself.

Decade after decade, technology slowly advanced, particularly in the realm of spectator experience. The constant development of broadcasting techniques now allows spectators to personalise their consumption of more than a thousand sporting events – up to the last detail. From choosing your camera angles to follow one player, through to selecting the voice of the virtual commentator – the possibilities are endless. What better experience than watching a World

Cup Final hosted by a synthetic Pele, with whom you can also interact, taking advantage of the latest advances in voice synthesis and artificial intelligence? The invention and improvement of neural implants, and the possibility of faithfully interpreting the flow of information that animates our brain, have allowed spectators to add to their arsenal of available camera angles a direct view from the athlete's eyes. Already since 2040, we no longer just watch a sporting event, we experience it from the inside. These improvements also allow for the increasingly intensive collection of physical data, which can be interpreted by algorithms that not only provide detailed analyses in real time but will also make it possible to refine the accuracy of predictions.

These increasingly accurate predictions have led to a real crisis in the world of sports and a profound reflection on its intrinsically unpredictable nature. Is a football or tennis match still as exciting as ever when the outcome can be predicted with near certainty, even before the players have set foot on the pitch? Betting platforms have feared the worst: what interest would there be in betting if you already knew the result from the start? That was until the 2043 World Cup where South Korea defeated the German team despite the predictions. The reason? Some of the forwards of the South Korean team were wearing cybernetic prosthetics to limit their muscular fatigue. Since then, the world of sports started to reinvent itself by gradually allowing the use of cybernetic devices, such as physical augmentations, genetic engineering or the injection of skills via implants. It is no longer a question of developing more efficient equipment, but of improving the athlete directly. This new situation has quickly had a twofold impact: on the one hand, it has led to a re-evaluation of the existing rules and adapted them to the new paradigm, thereby maintaining a sufficient degree of uncertainty to generate excitement and ensure the safety of these new augmented athletes. On the other hand, it has given birth to new sports categories – such as cyborg football – and even new sports in general, more in line with the unprecedented capacities of these improved athletes.

A whole new stage has been reached when an ambitious entrepreneur offered the public an even more intense immersion through close interaction with these athletes via their implants. And while the possibility of physical control (and therefore human drones) will soon be officially banned worldwide, this revolutionary promiscuity allows a privileged (and wealthy) audience to influence the course of a match to some extent, by giving strategic advice and instructions directly into an athlete's cortex.

The creation a few years later of the first androids – physical humanoid avatars inseparable from their models – has recently made it possible to get over this ethical limit and offer a formerly passive public the chance to control these prefabricated athletes in their entirety. This major innovation makes it possible, for example, for a quadriplegic to participate in a rugby game by proxy, with an experience similar in every respect to that of playing the sport at a high level. As android sports have become increasingly popular, specific championships have been created such as the Andro-VI Nations for rugby or the Android Ultra Soccer World Cup, just to name a few of them. In 2051, almost every augmented human sport has its android version. Aquatic sports are a bit behind in this respect as these bots must be improved for better underwater control. 2044 was the first year where the Olympic Games – in Addis Ababa – had four editions: historical edition, meaning the athletes weren't artificially augmented, Paralympic edition, augmented edition and android edition. But over the years, more and more Olympic and Paralympic athletes were using artificial augmentations. From a sniper aim eye – mostly used in archery, shooting and biathlon – to genetical mutation tripling the anabolism speed to recover faster in most sports, less and less athletes were participating in the 'historical' Olympics and Paralympic editions. Thanks to genetic and technological augmentation, Paralympic athletes have quickly been able to compete in the same categories as other augmented athletes. So much so that next year, 2052's Ulaanbaatar will be the first edition with only an augmented and android edition. New, strict regulations are applying to these augmented athletes to regulate these augmentations, specific to each discipline.

The arrival of the 'androbots' has unfortunately also paved the way to increasingly extreme sports, devoid of any moral or humanistic considerations. The famous 'duels to the death' between machines 'telepathically' piloted by humans – at first clandestine and illegal – has grown so much in popularity that it ended up becoming part of the sporting landscape (even if they still cause a lot of controversy today).

Yet, the androbots have given an even greater boost to e-sports, with a bigger following than augmented and android sports combined. League of Legends and Overwatch online competitions remain very popular for e-sport hardcore fans, but the League of Droids and Cyberwatch tournaments are on a whole new level. The last League of Droids live tournament that happened in February 2051 attracted more than three million simultaneous spectators.

And what a show it was! The best of the best players took control of the best champions in a crazy, pyrotechnical show that left everyone speechless. Not to be outdone, at the Cyberwatch tournament closing the Android Ultra Soccer World Cup, they're expecting 2,5 million spectators in the metaverse, with priority given to the participants of the Cup and 10 million spectators on Twitch – which thanks to 3D retransmission will give a good idea of how incredible the real show will be. All these changes have created a complete convergence between sports and e-sports. They have now become insepa-rable. Especially when you compare android sports with live – real life – e-sport tournaments. The only difference lies in the fact that android sports are limited to traditional sports disciplines that would be part of the Olympic Games. Whether it's virtual skirmishes in a synthetic universe or the piloting of android drones through neural connections, the distinction between these different practices is becoming blurrier and blurrier.

High-level sports, the kind that galvanise crowds, have undergone drastic changes compared to what they used to be in 2021. However, their 'classic' counterparts, the ones involving non-augmented humans still remain, even if some may consider them obsolete. More popular at a local scale, 'classic' sports continue to appeal to those still looking for the ultimate in self-transcendence, without artifice.

For others, this profound transformation of sport through technology has not only made it more spectacular, and contributed to its revival, but has also made it more accessible. Today, regardless of physical predisposition, it is no longer forbidden to dream of a gold medal at the Olympic Games.

MOVING THE GOALPOSTS ON THE VIRTUAL PLAYING FIELD

The sports world, hit hard by the Covid-19 pandemic, is going through an unprecedented crisis, the repercussions of which will force it to reinvent itself to stem the loss of revenue. Until now, professional sports have depended on financial income generated mainly by advertising and broadcasting, but now virtualisation could be the key to ensuring its survival. It is no coincidence that in 2020, the 24 Hours of Le Mans was first held online, via a driving simulator (RFactor 2). Similarly, 2020 has seen the advent of the first virtual cycling world's champions. These aren't just anomalies, but a harbinger of what is to come.

In addition to these new ways of experiencing sporting events, the emergence of new disciplines (and particularly e-sports) will gradually transform professional sports and force it to finally take advantage of technological advances, on the broadcasting side and on the field. Deloitte's report on the future of sports broadcasting says it clearly. Sports fans are looking forward to more technologically advanced experiences that allow them to watch their favourite game on-demand on their preferred channel with high quality images. They also want ultimate personalisation in their sports event consumption, which makes the idea of selecting your camera angle and your virtual commentator's voice a not so far-fetched perspective.

Nicholas Bostrom, professor at Oxford and a pioneer of simulation theory goes even further and says that game broadcasting could soon converge with VR simulations. To him, VR simulation could even open the way to completely artificial competitions with artificial athletes.

As for Androbots being the new athletes, Battlebot has been organising destructive robot battles since 1987. The robots are still quite basic, but a lot of research is being made to improve them in various disciplines. Japanese engineers have already developed a basketball robot specialised in precision, three-point shots. Robots are also central when it comes to drone races, and even have their own league. The Drone Racing League (DRL) has been organising huge events all over the world since 2015. One of the latest shows gathered more than 190,000 viewers.

With regard to augmented athletes, Thomas Frey, member of the Association of Professional Futurists, predicts that we might soon be able to create super-athletes by editing their genes before they are even born, thanks to CRISPR technology. Enhancing athletes' capabilities with implants or tech elements will be the next step.

Data will also play a huge part. It's highly probable that many metrics, including personal metrics from athletes, will be gathered and used internally for improvement purposes, but also externally for prediction purposes. The Wall Street Journal forecasts that fans would then use these data for betting purposes, which raises ethical questions. But considering that in 2020 and 2021, we were expecting public communication on athletes' Covid-19 test results, the same principle applies.

The crisis of 2020 sounded the alarm bell, initiating a profound rethink of professional sports for decades to come. And as this is certainly not the last time that the planet will confine itself to stop the proliferation of a virus, it is certain that this mutation of the sports world will accelerate. In the long run, freed from physical, technological and even ethical constraints, the only limit to sports will be the imagination.

PROBABILITY

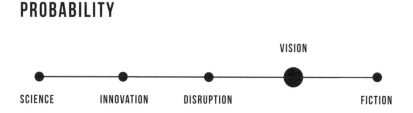

VISION

SCIENCE — INNOVATION — DISRUPTION — FICTION

The sports environment will undoubtedly change hugely in the coming decades. Some e-sport games could even become official Olympic disciplines before 2030, showing a convergence between 'traditional sports' and their online version. An experience like Tom's might actually be plausible, in a virtual stadium with Androbots being the stars of the game.

Data will help athletes to improve and perform better and better. However, as Nicholas Borstrom said, it's unlikely that this data will be of any help to spectators who would like to bet. And if you could, it would indeed reduce interest in the game.

In the future, traditional sports will probably still have their place in the heart of stadium lovers – to whom it would be inconceivable to have the same experience online. But it's also likely that augmentation is going to make sport even more entertaining for spectators that want more spectacle. Virtualisation will make them more and more accessible – both in terms of viewing and participating – pushing back the limits of what's possible. Both will probably cohabit the future. The question is: in what format and with which types of athletes? Bring on 2051!

BEYOND
THE BLUE

Marcelle wakes up feeling peckish. Not that she has any reason to be ravenous, with her bespoke nutritional supplements. It's just that today is a special day and she's going to eat actual food to celebrate the arrival of her twin sister, Mignonne, to the Basin. She checks the time – 5 am. Outside her small pod window, it's the usual, hazy view of the ocean floor, twenty metres below the surface. Every so often, a flash of light illuminates a tangle of structures in front of her window. Mussel socks and scallop lanterns float behind a sphere in the distance. It's attached to a greenhouse dome growing watermelons in saltwater. The light source: a swarm of tiny firefly squid, gene-edited to glow brighter if toxins in the water are detected.

These are just some of the sea creatures lured closer to the colony to help illuminate the exterior with their natural bioluminescence. It reminds Marcelle of her dinner menu. She needs to get ready in time for her first appointment – stopping by her friend, who has found some special goodies for her. After months of smoothies and kelp salads, she is really looking forward to the cultured meat schnitzels her friend has sourced, which she'll serve with a side of sea asparagus French fries, curried kelp and dehydrated corn pops. They'll wash it all down with a 'headache light' Sauvignon Blanc.

She hasn't seen her sister in over two years, since Marcelle and her partner Luc won a lottery to move to the island section of Neptopia. Situated 50 miles off the coast of Darwin, Australia, Neptopia has slowly come together since the 2040s as a place where like-minded seasplorers congregate to live a libertarian life. The enormous ocean farm has been in existence since the 2030s, but it was only when a famous Bitcoin billionaire decided to build a complete underwater city that things started to come together.

Neptopia means different things to different people. For Marcelle and Luc, it's a way to live a more meaningful life and see if they can tolerate being under-water for that long. There is a rumour of another Neptopia colony spawning off the coast of the Soloman Islands, where one can buy a 'sea smallholding' and cultivate a specific new type of microalgae used in deep ocean farming. In return for a harvest, residents are paid in the cryptocurrency Seacoin, and are

allocated special visas to live on any of the five micronations around the world.
There are only a handful of hairdressers in the ocean colony and Marcelle won
some extra points for being an English tutor, hairstylist and a pet groomer to
the wealthy Neptopians who were allowed to take small pooches with them.
After 14 months spent working on the island, the couple heard of someone who
had passed away in the Basin. It was finally time for them to move down into
the underwater nucleus for their six-month 'seabattical'. Luc works in transport
– ferrying passengers on highspeed hovercrafts from the mainland to the
island, delivering algae harvests to visiting ferries, and supervising a fleet of
autonomous, water purifying pods. Thank goodness for his great connections
because he was able to pull some strings to get Mignonne below (ocean)
surface in time for the twins' 40th birthday.

SEASTEADING: THE PERMANENT LIFE AQUATIC

'What if we could live underwater?' must be one of the most fun futuristic
games to play. With two-thirds of the earth covered in ocean, it's only natural
for futurists to consider a full-time aquatic life. Firstly, think of the design
opportunities for building a new, utopian world – especially when it comes to
sustainability. What if you could build a city around people, not cars? What if
every house, school, shop and factory could be designed to work in synergy
with one another, grow food, manage its own waste and generate energy?
Imagine upping the anchors and moving your city wherever you want it? Not
having to worry about land appropriation, even classism. With affordable
housing levelling the playing field, everyone starts from scratch.

But much as one can fantasise about libertarian societies, beautiful coral views,
and bathing in ocean water, rising sea levels are hugely concerning. By 2050,
90% of the world's largest cities will be exposed to rising seas. Currently,
around 40% of the global population lives within 100 kilometres of the coast.
These coastal cities are growing faster than they can grow infrastructure.

Shanghai, Mumbai and Miami are just some of the 570+ cities expected to be
mostly underwater by 2050. Cities that straddle lakes, sit on stilts, float from

harbours out to sea, or even exist entirely in the ocean, will be commonplace. We'll continue to see ingenious ways in which countries will deal with flooding. There will be 'sponge' cities like China's, massive sea walls like Jakarta's, and living water cities such as New Orleans. Some will pour giant amounts of sand into the water (witness Hong Kong); others will completely relocate their people like the Republic of Kiribati. This independent island nation in the central Pacific Ocean bought climate-refuge land in Fiji to house its Micronesian nation when (not if) it becomes permanently flooded.

Managed retreats, climate refugees and seasteading will be hot topics, as an answer to rising sea levels, over-population, farmland, or renewable energy sources. Seasteading – floating 'micro-countries' with 'start-up governments' – will give forward thinkers ways to experiment with new ideas that current governments are reluctant to try. These are some of the ideas floated (if you'll permit the pun) by The Seasteading Institute – founded by anarcho-capitalist Patri Friedman and initially funded by tech billionaire, Peter Thiel. Before Thiel let his dream go, he wrote in a 2009 essay how a new, libertarian society could be built on one of three technologies: cyberspace, outer space or seasteading – 'an escape from politics in all its forms'. The Institute sees seasteading as a very Silicon Valley way of thinking – that even governments can be 'hacked' and that, with nearly half of the world's surface an unclaimed, blank slate, they can create a platform to try out new nations and new ways of governing.

While some call the Institute's ideas elitist pies-in-the-sky (even 'tech colonialism'), the coronavirus pandemic has just fuelled the desire of libertarian groups to build autonomous new societies. Back in 2017, French Polynesia was on board with the idea, signing a historic agreement to develop the first floating city in a lagoon on the island of Tahiti. It ticked all the Silicon Valley boxes: exclusive, just an eight-hour flight from Los Angeles, tax haven, more or less self-governed, fibre to Hawaii, and spectacular surroundings. French Polynesia also has the world's largest exclusive economic zone (an area of sea stretching 200 nautical miles from a territory's coastline). In the end, there wasn't enough in the deal for Polynesians to go through with it. They don't pay tax anyway, they live in paradise, and they don't need people spoiling their view.

The Maldives, an archipelago of 25 low-lying coral atolls in the Indian Ocean and the lowest-lying nation in the world, is building the first-ever 'island city' ten minutes by boat from the capital Malé. Homes will float on a flexible grid across a 200-hectare, warm-water lagoon. Along with a ring of barrier islands, giant new coral reefs will be grown to act as water breakers. As the former Maldivian president said, when you can't stop the waves, 'you rise with them'. Designed by a world-renowned leader in floating infrastructure, Dutch Docklands, the company is also testing the technology in the Netherlands.

ADVANCE, NOT RETREAT

In 2019, the UN-Habitat – a United Nations programme for sustainable urban development – hosted a roundtable of architects, designers, academics and entrepreneurs to discuss the viability of floating cities as a solution to climate change and affordable housing. Hosted together with Oceanix, the MIT Center for Ocean Engineering, and the U.S.-based Explorers Club, this was where the concept idea of Oceanix City was first introduced. Designed by Big Tech darling Bjarke Ingels, this hurricane-resistant, zero-waste city will be made up of 4.5-acre hexagonal, floating islands that each house 300 people. It can be put together in infinite configurations: six islands form a village, and six villages form a small city of 10,800 people across 75 hectares.

This modular city is designed to house a man-made ecosystem with a circular economy. It can withstand severe weather conditions, produce its own power and plant-based food, create drinking water and handle waste disposal. Bio-rock will be used to make robust, artificial reefs for corals to grow. Food waste will be converted to energy and compost in community gardens, single-use packaging will be eliminated, and sewage will be treated in algae ponds. All vehicles (ferries, scooters, etc.) will be electric, of course. Fresh water will be supplied via the latest water vapour distillation technology, atmospheric water generators and rain harvesting systems. Energy will come from wind turbines, algae bioreactors, solar panels and wave energy converters. Food – from sea greens and fish to fruit and vegetables – will be grown in greenhouses, vertical and aquaponic farms and 3D ocean farming (like vertical farming, but underwater).

'Advance instead of retreat' is what drives Oceanix founder Marc Collins Chen, who was the minister of tourism of French Polynesia in the early 2000s. Chen was also involved in the French Polynesian ocean city that didn't materialise. When sea levels rise, retreat will be possible for places like Miami or Bangladesh. Some might go for a hybrid land cum floating city. Others, like Indonesia, are already making provisions for this flooding. Before the pandemic struck, Indonesia had firm plans to move its capital city of Jakarta to the island of Borneo. Speaking from experience, Chen is an advocate of pushing cities beyond the edge of the water. Managed retreats will definitely become part – or wholly – floating neighbourhoods, where roads become canals.

Oceanix is happening – already funded by a (secret) private venture capital firm and with the French company Bouygues Construction on board. Where it will be remains to be seen, or if it can be the solution to a global population that is fast running out of space. As the climate crisis worsens, more than one billion people will live in countries with insufficient infrastructure to withstand sea-level rise by 2050, according to the Institute for Economics and Peace. At this rate, it would take over 9,000 Oceanix cities to rehome these projected climate refugees.

If you're worried about how such a vast, floating structure will cope in high seas with storms and 80-foot waves, experts in naval architecture and marine engineering are already on top of it. MARIN (Maritime Research Institute Netherlands) has tested an innovative concept for a floating mega island that could provide a living and working space at sea for developing, generating, storing, and maintaining renewable energy, including offshore wind. These modular, three-sided islands can work for any size and, eventually, in any kind of weather. MARIN has already tested this structure in its test pool (one of the largest indoor pools in the world), pummelling it with the equivalent of 82-foot waves. With a flexible, wave-facing part of the island cushioning the blows, no crests crashed over it. Whatever is used in future, whether it's hinged triangles, spars or pontoons, it won't be long until a failsafe solution is on the table.

THE ULTIMATE OCEAN VIEWS

Practicalities aside, let's dive into the fun part: what these floating buildings and rooms could look like. If you want to own a piece of that fish tank feeling, you've missed the boat with Dubai's stunning Floating Seahorse Villas. All 133 of these have sold but you can probably rent one to experience the brilliance of floor-to-ceiling glass as you admire marine life and colourful coral reefs from your bedroom. Even at night, you can watch the reefs come alive with nocturnal marine activity. Each villa has two storeys with four en-suite bedrooms, indoor and outdoor living areas, a roof with an infinity swimming pool and a glass floor so you can see views of the water below. There will also be a kitchen, living room and two staff quarters.

Water Discus Hotels, the prototype by Polish company Deep Ocean Technology, feature a modular design comprised of two discs – an underwater and above-water one – that resemble saucers. Each hotel will be situated ten metres underwater to make the most of sunlight, and will contain twenty-odd rooms, a bar and a restaurant with views of surrounding coral reefs. There has been talk of construction in Dubai, Singapore and the Maldives, but nothing has materialised yet.

Poseidon Undersea Resort is another yet-to-be-built underwater hotel that will have submarine-style capsules and even an underwater wedding chapel forty feet underwater. Spread across two hectares, nearly 70% of the suites' surface areas will be transparent, providing spectacular views of the undersea world.

Other underwater rooms exist at the InterContinental Shimao Wonderland in Shanghai, Reefsuites on Australia's Great Barrier Reef, Hilton's Conrad Rangali Island Resort in the Maldives, and the Underwater Suites at Atlantis, The Palm in Dubai. If luxury resort living is not your thing (or in your budget), perhaps the privacy of an Anthenea pod is more your cup of kelp. These eco-friendly mobile homes are completely off the grid, with a futuristic design inspired by the James Bond film, The Spy Who Loved Me. Guests can enjoy panoramic, 360-degree views from their round beds or peek at sea life through glass panes. The best part? These pods can be sailed around the world, while black and greywater are treated on-board, releasing clean water back into the ocean. On an even more modest, yet perfectly achievable scale, is the Makoko Floating System (MFS) by NLÉ – a very simple way to build on the water by hand.

It is a prefabricated, modular, floating A-frame structure made from sustainable timber. It can be locally produced, assembled and disassembled – and quickly too. Designed by Nigerian architect Kunlé Adeyemi, it has evolved from a floating school to being deployed in five countries across three continents. Adeyemi created this design in an effort to reimagine life in Makoko – the floating slum in Lagos that is sometimes called the 'Venice of Africa'. It's inhabited by around 250 000 people who not only live on water but also depend on it for their livelihood. It will be interesting to see how this prototype evolves and if it meets Adeyemi's vision of African Water Cities.

Perhaps, instead of building new water cities, we'll find a way to inhabit the plastic islands that already exist. In 2021, a new study found that the Great Pacific Garbage Patch floating in the Pacific Ocean between California and Hawaii is 16 times bigger than was previously thought. It's said to be made up of enough fishing nets, plastic containers, ropes and microplastic to fill 500 jumbo jets. The Ocean Cleanup is making good headway in catching this environmental disaster, with its fleet of long, floating barriers that act as an artificial coastline. Using winds, waves and currents to passively catch and concentrate the plastic, they slowly drift along the 1.6 million square kilometres of this trash vortex – three times the size of France.

But what if these islands can be turned into liveable islands? That's exactly what architect Margot Krasojević is proposing. Her concept for a luxurious, 75-room hotel will use large, plastic-filled bags woven together as an anchor to the ocean floor. These bags will be weighed down with silt and sand to make the structure stable. There will be artificial, mangrove-like 'roots' to trap sediment and act as a flood defence by sucking up water to inflate when needed – almost like a lifejacket. With funding already in place, this could be a new form of eco-tourism where, as you sip your pina coladas on the pool deck, your floating hotel filters and saves the ocean.

Architect Ramon Knoester's Recycled Island also plans to use plastic from 'trash island' to create a completely self-sufficient island the size of Hawaii. It will support its own agriculture and derive its power from solar and wave energy. When completed, it could house 500 000 residents. Keen to see it happen? The project is still floating around on Kickstarter, so it's ripe for the taking.

LIVING LIKE THE AQUANAUTS

Staying a few nights in an underwater hotel or zipping around in a pod for a week is one thing but living underwater for extended periods of time needs to consider every possibility, from security and bad weather to food and water. So is it really possible? Absolutely, according to Ian Koblick. As a pioneer of ocean exploration since the 1960s, he was one of the first aquanauts – those who not only work underwater but live underwater too.

Ian designed and operated *La Chalupa* in the 1970s – the most advanced undersea lab in the world in Key Largo, Florida. In 1986, it was converted into Jules' Undersea Lodge, at the time the world's only undersea hotel, and operating an environmental education centre. It was at this lodge that two scientists broke the record for the longest time spent living in an underwater, fixed habitat. They spent 73 days, 2 hours and 34 minutes living 7.31 m (24 ft) underwater. As someone who has often spent two to three weeks underwater, Koblick thinks longer periods for larger groups are completely possible. Neither he nor his colleagues ever experienced any ill-effects from living below the surface, so he thinks up to six months would be feasible. According to a 2013 interview, he doesn't see any technological hurdles to making this happen – it's only a matter of money. Okay, and things like emergency evacuation systems, controlling air supply and humidity, not to mention human adaptation to extreme environments (read: psychological effects).

This is something closely monitored at 'the world's gateway to inner space' – the Aquarius Reef Base in Florida and the training ground of NEEMO – NASA's Extreme Environment Mission Operations. Groups of astronauts, engineers and scientists are sent to the world's only operational and habitable undersea saturated environment to prepare for space exploration. Space hopefuls fulfil a series of simulation missions, like practicing space walking. If you're wondering what 'saturated' means, it's what happens after spending 24 hours underwater at any depth, when the human body becomes saturated with dissolved gas. In essence, this means that aquanauts can remain underwater for an indefinite amount of time.

Creating underwater habitats has been a lifelong dream of Koblick and is also the focus of Open Sailing – an international community of architects, engineers, inventors and scientists working on Open Sailing technologies to explore and study the oceans. The project began as an apocalyptic design response unit but has evolved into a collective that wants to build an International Ocean Station (like the International Space Station but for ocean living). Some of these technologies included Nomadic Ecosystem (mobile aquaculture to sustain long-term life at sea), Life Cable (unified standard for energy, water and waste), the Open Politics think tank on how to organise a new oceanic urban structure, and Openet.org – a purely civilian Internet moderated only by its users. The idea of Instinctive Architecture meant that the whole structure would behave like a super-organism, reconfiguring itself according to weather conditions. For example, in stormy weather it could shrink to protect itself from waves and wind, and in calm seas it would expand in different directions to have more space for fish or seaweed farming.

Someone who is making this dream come true is Jacques Cousteau's grandson Fabien, who is going to build a space station of the sea named after the prophetic sea god, Proteus. The station will be located at a depth of 60 feet, in a biodiverse, Marine Protected Area off the coast of Curaçao. It will be ten times the size of the Aquarius Reef Base, and will contain living quarters, research laboratories, medical bays and bathrooms. Proteus will also include an underwater greenhouse so that inhabitants can grow fresh plant life for food. One of the main aims of Proteus is to increase our knowledge of the ocean seafloor through high-resolution, 3D mapping. About 80% of the ocean hasn't been explored yet, so this will be a critical tool for protecting and tracking marine life, regulating underwater exploration, assisting with rescue missions and predicting natural disasters like tsunamis. The better we can map the oceans, the better we'll understand ocean circulation, weather systems, sea-level rise, and climate change. Eventually, the hope is for a whole network of Proteus bases, situated throughout the ocean.

DEEP OCEAN MARICULTURE AND OCEAN CROPS

If we were to live on or underwater for long periods of time, food cultivation is obviously high on the agenda. Global fish demand is set to double by 2050, with experts predicting that aquaculture (farmed fish) will be responsible for up to 90% of this supply. While farming fish on land is well-established, and growing filter feeders like mussels and oysters is very doable, it's the carnivore fish like tuna that are more challenging to farm. Farms will increasingly look at fishmeal alternatives like insect, algae and plant-based aquafeeds. If we're going to feed nine or ten billion people in 2050, we can't do that by grinding sardines and anchovies forever. With some of the highest concentrations of EPA and DHA polyunsaturated fatty acids of any fish species, these fish should be our food, not fish food.

Looking at mariculture innovations, this could actually drive ocean living faster than we think. There is a 'blue rush' going on right now to exploit the vast, unclaimed ocean resources. Whereas marine fish farming usually happens close to the coast in sheltered bays, there is a move towards farming fish much deeper in the ocean. Take, for example, Open Blue, which runs the largest open-ocean fish farm in the world. Located 11km off the coast of Panama, this cobia farm has 22 pens that can produce 1,200 tons of fish annually with no effect on the environment.

SalMar's Smart Fish Farm will be significantly larger than its Ocean Farm 1, with a production capacity of 23,000 tons round weight. This super-cage will exist in an area of open sea between 30 and 70 miles off the Norwegian coast. It's the first time someone has applied for approval of a site for aquaculture in the open sea (with the Norwegian Directorate of Fisheries). It will also take salmon fishing deeper into the ocean than ever before, with a vast structure that's able to withstand a 100-year storm.

Going even deeper: Ocean Era's floating aquapods have trialled 'over-the-horizon' fishing of kampachi, off the coast of Hawaii. Two thousand of these sashimi-grade fish were put inside the sphere and attached to a remotely controlled, unmanned feed barge which would then drift up to 75 miles off the coast. Technicians ran the farm remotely, from a smartphone or tablet, only visiting the site once a week to top up the feed or fuel the generator.

The company will also trial a submersible growing platform for microalgae, moored approximately 120 metres below the ocean. This means there will be no impact on water quality, coral reefs, or dolphin resting activity.

When we look to the future of ocean food, the word that always crops up is seaweed. Or, as its advocates like to call it, sea greens. With over 10 000 edible plants in the ocean, expect to see these green superfoods incorporated into our daily meals just like rocket, spinach and kale. Sweet sea moss smoothies, kelp carrot cake, pickled kelp, sugared kelp, seaweed mash... your culinary lexicon will get to know words like wakame, dulse, arame and hijiki. These restorative crops are cheap to grow and could help us in the fight against climate change and ocean pollution.

Making regenerative ocean farming accessible to all is Bren Smith's pioneering, 3D vertical farming method. Through his company Greenwave, he is sharing his revolutionary, open source polyculture farming system with the world. It can grow a mix of seaweeds and shellfish with zero inputs – making it the most sustainable form of food production on the planet. It also sequesters carbon and rebuilds reef ecosystems. All you need is a boat, about 20 acres of ocean and $20-50K in start-up cash.

Seaweed, mussels, clams, oysters and scallops all grow together happily in this vertical tunnel – even in waters that are dying from acidification. Now consider the vast job creation potential seaweed farming could have and it's easy to imagine future entrepreneurs living permanent ocean lives. According to a World Bank study, farming seaweeds in just 0.1% of the world's oceans (about 100 million acres) could create 50 million jobs.

Fish aside, what about growing fresh produce in the ocean? Ocean crops offer an exciting new frontier for agriculture. Companies like Agrisea are testing how rice can grow in saline conditions and is confident that large floating crop islands will be a common sight in thirty years' time. Researchers from Nottingham University have come up with Floating Ocean Farms – floating containers where crops can be grown using hydroponics or aeroponics. Containers would be lit by LEDs to maximise photosynthesis, while power would come from offshore wind turbines, wave or tidal power.

Off the coast of Italy, the Nemo's Garden Project by Italian ocean diving company, Ocean Reef Group, has been growing strawberries, orchids, basil and lettuce in pods on the ocean floor since 2012. They've recently teamed up with Siemens to further study and improve this technology to finalise its industrialisation as a sustainable food alternative. If we can be even more adventurous, we could be eating salt-loving sea vegetables like sea asparagus, saltwort and sea purslane. Rich in nutrients, these veggies grow in as little as eight to ten weeks.

ENERGY ON THE OCEAN

The final piece of the ocean living puzzle: how would these underwater cities, farms and factories be powered? Naturally, solar power is the first logical idea. After all, the sun shines just as much on the water as it does on land; plus, the seawater can help cool the solar panel technology. One could take inspiration from the world's largest floating solar farm on Indonesia's island of Batam. Built on an area of 1,600 hectares, it will have an expected output of 2.2 GW. It will also have the largest energy storage system with a capacity of over 4000 MW. Or we could look at Norway's SINTEC who will also test a solar farm at sea. In an effort to cope with stormy seas, researchers have tested an anchoring system that will give the installations enough freedom to cope with large waves.

But perhaps we need to think much bigger than wave, sun and wind energy. How will we power farms deep under the ocean where it's dark? Plankton power could be the answer – especially to power oceanographic sensors that constantly need replacing. The US Naval Research Laboratory created OSCAR (Ocean Sediment Carbon Aerobic Reactor) that taps into a natural voltage gradient created by a pair of chemical reactions happening on the seabed.

And, who knows, maybe by 2050 scientists will have figured out how to harness the endless energy from underwater volcanoes. Most of the Earth's volcanic activity takes place several kilometres underwater. Scientists have only recently monitored how far these underwater explosions (known as mega-plumes) stretch, but measuring the ripple effect showed enough energy to power an entire continent.

PROBABILITY

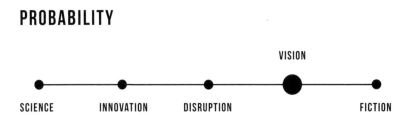

VISION

SCIENCE — INNOVATION — DISRUPTION — FICTION

What will push us into the ocean for good? The 'blue rush' for algae farming? Overheated cities? Power-hungry dictators? Some sort of natural disaster that makes the air unbreathable? Floating farms, ocean colonies and clusters of underwater micronations – it's only a matter of time before some version of these becomes the eighth continent, with a cool name like Oceania, Neptunia or Zealandia. Perhaps it will be a much larger version of Sealand – the world's tiniest (unofficial) sovereign entity. The original Sealand has been going since 1967, complete with its own anthem, currency, football team and hyper-secure data fortress.

Who wouldn't want their own country where you can do whatever you like, whether it's working a three-day week (or not work at all), enforce veganism or do as Liberland does, simply 'live and let live'? Perhaps it will be the promise of a simpler life, some peace and quiet, away from the maddening, mind-bending metaverse noise, basking in the ultimate cocoon life of perpetual daytime and magical, luminescent fireworks. As soon as it's possible, sign us up. To quote Jules Verne: 'The sea is everything.'

POSTCARDS
FROM TORPORVILLE

Tung's eyes slowly flutter open, adjusting to the dim lighting. He stifles a big yawn and stretches his legs. He's feeling rested, as if waking from a long afternoon nap, but with a slight sour taste in his mouth. He doesn't quite know where he is, but the custom-designed aromas blowing over his face remind him of his home on the Yellow River, so he feels right at home. Suddenly, a voice is speaking to him as the lights start to brighten. He notices that he's inside a padded pod, and shivering just slightly. 'Hi Tung, welcome back. You've successfully Hiberday-d for 180 days. Take a moment to breathe and relax as we bring your body temperature back to normal'.

Now it slowly comes back to him in waves of memory and relief. About seven months ago, one dark Friday, he sat down, hands in hair, looking at his mountain of debt. For years, he's had a successful tour company taking tourists into the Huang River Valley for thrilling adventure holidays. His point of difference: he was one of a handful of operators in the world still offering visitors the joy of petrol-driven vehicles. Nothing is as exhilarating as a noisy petrol or diesel car. People travel from far and wide to get behind the wheel of a 2020s-era Dodge Demon or turn-of-the-millennium BMW Z8. The waitlist to feel the roar of a Harley Davidson under your pelvis is months long. It's the ultimate bucket list holiday – reliving your childhood, driving faster than you're allowed anywhere on earth, and enjoying beautiful scenery. Some people want speed, others are just happy with the smells and sounds of a good old Mini Countryman.

It's a business that Tung built up over fifteen years, having to grease many palms to get hold of supercars and ordinary heroes from around the world. China has been slow to join the carbon zero mission but a string of natural disasters compelled it to make drastic changes. Since 2035, petrol or diesel cars were no longer manufactured. By 2045, all remaining petrol cars on the road were heavily taxed. Add to that all the other green taxes – anything from lighting your barbecue, to not properly recycling your old clothes, to how many children you have – and it became quite a tricky business model. Yet, Tung made it happen with smart environmental investments.

Everything was still fine money-wise until a devastating dust storm killed his business for a year. Insurance paid out but it was the final nail in the coffin.

He was about to file for bankruptcy when a government insider told him about a brand-new programme where you can save money by going 'off-grid' for a few months. The company takes care of it all – they shut down your home, look after your pets and use the savings of your environmental non-impact to pay off your debt. Like thousands of others in big Chinese cities, all you are required to do is to go to sleep in a pod for six months. It's totally safe and saves the country tonnes of carbon emissions. 'Like that Snowpiercer movie?' Tung asked. 'Yes', said the friend. 'But it's not a drawer, it's a comfy pod. And it's not a punishment, it's the number one way to make money right now! I would if I didn't have a baby on the way!'

And so, Tung told his family, said goodbye to his dog and travelled to Chongqing for his Hiberday stay. Six months of not worrying about bills to pay, six months of stress-free sleep.

HIBERDAY: YOUR WAY TO FINANCIAL FREEDOM

From 2050 and beyond, voluntary coma holidays will be a novel way to escape life for six months to a year. You'll do it to take part in paid experimental treatments, to 'rewire' your brain after a traumatic incident, or to undergo various gene therapies all at once for healing and anti-ageing purposes (not to mention saving a lot of time). Perhaps, for those who can afford it, it will be a way to combine an ultra-restful sabbatical with a lifetime brain download into the cloud and intense hair follicle stimulation. It could even be a way for countries under severe environmental pressure to cut down on the resources used by 'pausing' parts of their population.

Either way, being out of circulation for such a long time has many economic and environmental benefits too – something that a clever start-up like Hiberday will turn into a booming business. What happens to your home, your job, your bills? Who feeds your dog and waters your plants? Imagine the money you'll save not having to pay rent, buy groceries or pay for entertainment. What about all those sky-high carbon taxes you owe each month? Wrap up the idea of a slumber holiday with a clever emissions trading scheme and you have a winning business idea.

Why not go off the radar for a while if you can save some money for a home renovation or to send your child on a space field trip? Why not pay off your debt once and for all? Voluntary coma holidays will become as mainstream as August beach vacations for Europeans – without the stress of travel, the inevitable sunburn and screaming children.

Hiberday's concierge service will take care of it all. Its clever network calculates exactly who will be out of circulation and whose homes are available for holiday house swaps. It interfaces directly with government and debt agencies to help clients manage their finances, taxes and other commitments. Its recruitment division will find a replacement or work experience opportunity for someone else.

HUMAN HIBERNATION: CLOSER THAN WE THINK

Statis, hibernation, or suspended animation. Whatever you want to call it, humans have long been captivated by the fact that some animals can take extra-long naps with absolutely no side effects. Known as torpor, it's a state they enter where bodily functions are reduced to a minimum and they use fat stores for energy. Some animals like mice and hummingbirds enter torpor daily to preserve energy. Others, like hedgehogs and the Madagascan dwarf lemur, hibernate for months on end. Arctic ground squirrels (understandably) disappear for eight months of the year, dropping their basal metabolic rate by about 99%.

While these squirrels were obviously built for extreme cold, one can't help but wonder why, if bears and primates like lemurs can hibernate, can't humans do so too? The answer is that they can, in theory. Sort of. Therapeutic torpor has been around since the 1980s and has been a staple for critical care trauma patients in hospitals for almost twenty years. Search 'human hibernation' and you'll come across many survival stories – the most impressive being that of Mitsutaka Uchikoshi. Walking home down a mountain, he slipped, broke his pelvis and eventually lost consciousness, only to wake 24 days later. His body temperature fell to 22 degrees Celsius (normal is around 37), he barely had a pulse, lost a lot of blood and suffered multiple organ failures. Yet he somehow survived and recovered fully.

So, whether it's to solve questions surrounding ageing or deep-space travel, to alter metabolic rates for weight loss, or to preserve pulseless trauma victims during life-saving surgery, human suspended animation is not that far off at all. The Japanese government is investing close to $1bn in futuristic scientific projects over the next decade – which includes 'artificial hibernation' – with the hope to prolong the lifespan of its disappearing ageing population.

If a nine-month trip to Mars is to come true, NASA or the European Space Agency will have to perfect their RhinoChill method to lower astronauts' body temperatures to keep them in a sleep-like state during space travel. There are many ways to put human beings into a deep sleep state – either via a chemically-induced coma through a controlled dose of barbiturates, or through temperature-induced hibernation. This is where cryogenic processes preserve a person in a suspended state, by slowly lowering body temperature to the point where metabolism, heart rate, and respiration slow down completely.

RhinoChill works with tubes that shoot cooling liquid up the nose and into the base of the brain, which induces a hibernation-like state. While the astronauts are asleep, robots would administer intravenous sustenance and electrically stimulate their muscles to keep them fit and strong. One crew member would stay conscious while the others hibernated for two-week periods. This would be maintained on a rotational basis until they reached Mars.

If this all sounds a bit extreme, should we rather develop a drug that could drop a person's core temperature safely and help them into a bear-like torpor? Scientists from the University of Alaska Institute of Arctic Biology are working on exactly that. They've been studying arctic ground squirrels for years and are working on a drug that could turn down your thermostat, so to speak. It's already working reliably in rats, so they are in talks with the U.S. Food and Drug Administration about human testing.

Of course, for now, studies are focused on health conditions. According to the Institute, cooling body temperature can help to treat many inflammatory iseases. Temperature-modulating drugs could help reset metabolic pathways affected in obesity and diabetes. Cooling methods used for emergency preservation and resuscitation could in future be used for people suffering heart attacks, strokes or exposure to poison. Whatever can buy doctors a bit more time is very exciting work – Mars trip or no Mars trip.

The Alaskans aren't the only ones making waves in torpor research. Harvard Medical School neuroscientists have discovered a population of neurons in the hypothalamus of mice that controls hibernation-like behaviour, revealing for the first time the neural circuits that regulate this state. There is also the renowned cell biologist Mark Roth, who has discovered that by boosting levels of iodide in mice, pigs and rats, you can help them recuperate better after traumatic events than those who didn't receive the treatment. Further developed, this approach could help people safely come back from hibernation without damage from an oxidative burst (the body's 'freak out' reaction to stress).

PROBABILITY

VISION

SCIENCE   INNOVATION   DISRUPTION   FICTION

Putting humans into hibernation or a long-term coma? Very possible. Putting thousands of people in sleep pods? It's not the craziest idea. While we don't know the exact figures, medics from the University of Maryland School of Medicine have already placed at least one patient in suspended animation in a trial approved by the FDA. Doctors remove the patient's blood and replace it with ice-cold saline solution. The patient, technically dead at this point, would then be operated on, before having their blood restored and being warmed back up to the normal temperature of 37°C.

As Professor Samuel Tisherman, one of the pioneers in Emergency Preservation and Resuscitation (EPR) procedures, rightly says: 'We're not trying to freeze the dead, just buy enough time to save the living.' Whichever way this works out, we're in for an exciting (and chilly) ride.

MY BODY, THE HARD DRIVE

It's a big day for Cassandra, a gastronomic engineer from Helsinki. She is presenting a new recipe for a delicious jellyfish and pea-milk cheese sausage to a global fast-food chain. A species that thrives on pollution and climate change, jellyfish have been slowly choking the ocean – and given gastronomists much to experiment with lately. In a mostly meat- and fish-free world, Cassandra is hoping to take the junk food world by storm with her new creation.

She worked on her proposal late into the night – documenting images of ingredients, styling plates, and recording high-definition, slow-motion footage of the oozy, melting pea cheese inside a neon-green casing. As always, that is the mouth-watering clincher. It really is a thing of beauty. Now all that remains is to safely carry it with her on her 12-hour hydrogen-electric plane flight to New York. She can't risk having her idea stolen or leaked by IP cybercriminals working for competitors.

Cassandra plugs a small device into her camera with a circular sensor that painlessly punches into her neck with tiny needles. While she sips on her mushroom coffee, files are quickly transferred into her bloodstream and saved directly into her DNA. In about thirty minutes, her designs are safely encrypted and ready to take with her.

At the airport, a simple iris scan will access Cassandra's passport, travel and medical history – also stored in her body. She has just received her booster vaccine against Covid-37. A tiny subcutaneous pellet was implanted under her skin and will guarantee immunity for five years. On the plane, she uses a mobile neck patch to download some old photos from her childhood onto her tablet. She plans to revisit some of her favourite spots, so she is scanning through her photos as an 8-year-old to find her old home address.

When they fly over Newfoundland, her tablet pings. It's the photo she's been looking for. A sizzling summer day in Cobble Hill, Brooklyn. Cassandra biting into a hot dog, smiling into the camera, juices running down her cheeks. A perfect moment of meat-eating happiness, blissfully unaware that in 2050, Cobble Hill would be permanently flooded. An underwater island, swimming with memories.

WHEN YOU'RE PART OF THE DNA-OF-THINGS

Medical records, CCTV footage of your home, the video of your child's home birth, bank accounts and more. In 2051 we will no longer trust computers, services or clouds to store our most private information. We'll store that data in our own bodies, at an atomic level. That way, we know that everything we hold dear is carried with us and nobody can access it – unless we are physically present. No more threats from hackers, no more stressing about the cloud going down. We predict that this technology will become mandatory – including storing our ID cards and visas.

By removing the personal information of nearly 10 billion people from the cloud – finally completed in 2045 – the world will be saved from a certain data apocalypse. The first warning came in the early 2030s, when the unfortunate timing of a tsunami hitting the United Kingdom at the exact moment fires were raging in California brought down the cloud for eight days. In future, industries will rely on a combination of storage methods, from quantum computing to synthesised DNA storage. Genetically encoded digital data mixed into common manufacturing materials will become commonplace, so humans will be yet another vessel in the DNA-of-things (DoT).

FROM DNA STORAGE TO DNA COMPUTING

'Software is eating the world', Marc Andreessen famously said. But very soon, data will eat the world. It's no longer a question of 'will we run out of cloud space', but when. At this point in time, our digital universe is already made up of more than 44 zettabytes of data (1 zettabyte = 1 trillion gigabytes) – 90% of which was generated over the last few years alone. It is forecasted that we'll have around 572 zettabytes of data by 2030 (10 times more than today). If you use a model of exponential continuation, the world will have 50,000 to 500,000 zettabytes of data to deal with in 2050. To picture one zettabyte, think of as much information as there are grains of sand on all the world's beaches.

We are heading for a certain data apocalypse. We are generating data at breakneck speed, and making no effort to dispose of anything. Just imagine the catastrophe if all the data in our connected world is wiped out. Therefore, scientists are already experimenting with alternative means of storing data such as liquid-state, helium, glass and holographic storage. IBM is working flat out to make its atomic-scale storage economically manufacturable. But let's not forget about nature's ultracompact and stable storage medium that's been perfected over 3 billion years of evolution. DNA storage is nothing new. All the living things on planet Earth have been using it for millions of years. Every human, reptile and fish is born with a blueprint running through their veins that tells it how to grow, survive and fight disease. We're not only storing lots of data in our DNA, our cells are constantly accessing and computing it. What if we can harness that computing power?

DNA-based data storage systems will be able to hold and process massive amounts of data. With an information density that's millions of times better than conventional hard drives, a single gram of DNA is able to store up to 215 gigabytes. Scientists have stored many things in DNA already, from the book *War and Peace* and the entire 16GB of Wikipedia, to the Netflix series Biohackers. So far so good – they seem to have the encoding and decoding of binary data to and from synthesised strands of DNA down. The next stop? Creating a living library. In 2017, Harvard researchers used the CRISPR system to insert DNA encoded with photos and a GIF of a galloping horse into live E.coli bacteria. CRISPR is of course the revolutionary molecular tool that combines special proteins and RNA molecules to precisely cut and edit DNA. Even after multiple generations of bacterial growth, they were able to recover the GIF by sequencing bacterial genomes.

DNA storing was always perceived as slow and expensive, but experts are already predicting that, with the right investment, costs could drop to $100 per terabyte by 2025. Everyone, from Microsoft to a whole host of start-ups

(Catalog Technologies, Iridia, Helixworks Technologies and Cache DNA) is taking part in the race to build commercially viable, DNA data storage technology.

Cache DNA's aim is to provide a low-cost platform to store nucleic acids that are mission-critical to a number of areas, such as viral detection, ecological conservation and forensic analysis. In other words, should we land up with an atom bomb or meteorite scenario, these kinds of back-ups will save us from going back into the technology middle-ages. Meanwhile, Catalog is likening its system to a printing press that will synthesise batches of many different kinds of short DNA sequences. As you would scramble letters to form words, the original binary data is encoded by stitching together 'DNA letters' into billions of possible words. These sequences are then stored in teeny tiny powder pellets – ready for when you want it 'read back', even 100 years from now. Once we have these nanoparticles, just imagine where they'll be inserted. This DoT storage process could hide useful data in everyday objects (steganography) or manufacture objects containing their own blueprint.

PROBABILITY

VISION

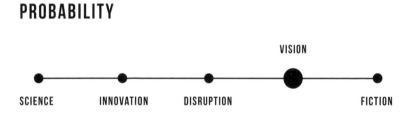

SCIENCE   INNOVATION   DISRUPTION   FICTION

Thus far, DNA-based data encoding is likely to be a long-term storage option for 'cold data' – archiving vast amounts of information that don't have to be regularly accessed. The prediction is that so-called 'hot data' – the very personal information we mentioned before – will be captured in other formats.

But who knows? Perhaps technology will advance faster than we think. And maybe, just maybe, instead of storing our most private data on data stickies or tiny glass coins, we'll store it directly in our own bodies. Then again, maybe all Cassandra needed to do to protect her IP was to swallow biocompatible silica beads containing her files and make a quick trip to the ladies room before her meeting...

UNIVERSAL
MIND CONTROL

December, 2051. As usual, Sarah is quite late for the family's New Year's Eve dinner. She's always racing against time to meet a work deadline. Luckily, what in other times might have been a complicated logistical ordeal is now managed with childlike simplicity. Sarah activates her Subcutaneous Personal Assistant (SCPA), a discreet little chip located just under the skin on her forearm. This feature allows her thoughts to be transposed into commands. Her report will be 'typed' for her and sent to her team in the Sydney office well before they start the workday, allowing Sarah to make up some valuable time while showering.

While she brushes her hair, Sarah frowns at her reflection in the mirror. She's looking pale. Too many hours spent at her desk recently. The mirror instantly redirects her to a quick makeup tutorial, which helps put just the right amount of colour into her cheeks to give her a healthy glow.

As a final preparation, Sarah puts on an outfit pre-selected for her by her AI personal assistant (according to the context of the dinner and the mood of the day, of course). She has been feeling slightly premenstrual recently – with a bit of water retention – and is therefore relieved to see that the outfit chosen is flattering around the waist.

As she puts on her shoes – and simply by deciding it's almost time to leave – she is simultaneously ordering an autonomous vehicle to take her to her destination. No need to ask her where she wants to go: the 'virtual taximan' has already accessed her diary and address book, knows where the journey will take her and can already plan the ideal route.

On the way, Sarah remembers it's her best friend Amber's birthday. A few seconds later, Amber pops her a message from New York to thank her for the gorgeous flowers. She takes the opportunity to order a last-minute bottle of wine, which will be delivered by drone directly to her parents' home. She didn't even have to specify the grape variety; the shopping assistant spontaneously asked about the meal prepared by Sarah's mother and chose the wine that would perfectly match the main course. In a few more minutes, she will reach her destination.

The route chosen by her vehicle decides, seemingly on the spur of the moment, to take her past her childhood home. This is of course, no accident. The house is now for sale, and Sarah has been feeling nostalgic about it for the entire day – tempted to buy it herself. The garden would be perfect for her son, Seb. She has been speculating how the neighbourhood looks these days. And optimised by the data transmitted in real-time by users of the surrounding roads, the car gives her a glimpse of what has been happening recently, which restaurants are still in business, and how the park looks lately. Thankfully, it's not far out of her route and the car, making up time, still allows her to catch up the few minutes she lost earlier. On the way, Sarah enjoys a quick game of Fortnite 2050, checks her emails, responds to her team's questions regarding her report, and replays moments from Seb's last basketball game in subjective view. Satisfied with her child's performance, she mentally swipes this retrospective into the family virtual space, so that she can proudly share it with the other dinner guests.

When she reaches her parents' house, her mother greets her at the door, looking worried. She explains that Sarah's younger sister has once again brought home an unexpected guest, and she is concerned that the supper will not be enough to feed everyone. Sarah smiles and greets her mother. A few seconds later, the drone bearing the bottle of wine arrives, with an extra pack of groceries to match the dinner ingredients. Problem solved.

The evening promises to be beautiful.

THE AGE OF THE UBERNET

Although the notion of 'augmented humans' is not new (people were already wearing glasses in ancient times), the idea of a fusion between human beings and machines only takes off for real at the beginning of the 21st century. After a transitional period during which connectivity between people existed exclusively through mobile devices, which were becoming smaller and smaller, the creation of neural implants and the advent of nanotechnology made it possible to integrate these interfaces directly into our brain. This has replaced the 'wearables' and augmented reality of yesteryear with a permanent, internalised state of connection, transforming the human being into a real, connected hub. This convergence between the digital and physical worlds, coupled with a considerable expansion of the Internet, has led to the creation of what is now called the Ubernet. This vast network, which will lead to the complete digitisation of our relationships, collects and processes the personal data of billions of people every second, and through dedicated algorithms, designs real-time applications to improve their life experience.

A global interconnection and an endless stream of information, analysed and processed by dedicated software embedded directly in our synapses, have gradually enabled us to perceive the physical world with greater acuity. We no longer need a device to connect to the rest of the world; yesterday's Internet is now part of us, and gives us instant access, without any external artifice, to an almost unlimited source of information. The human is no more. Welcome, connected super-human.

Now, in the mid-21st century, we consume music, movies, TV series, or sports events via nanometric devices that interface directly with our optical and auditory nerves; we can easily access our memories, delete those we consider superfluous or painful, share pleasant anecdotes with our loved ones, all in the form of raw moments that we can revisit in an emotional way, and no longer only through verbal communication.

A large part of our professional and human interactions now goes through the Ubernet, passing from brain to brain, intuitively, without any regard to the distance that separates them. This new promiscuity has, of course, led to a fundamental change in the way we now conceive of our relationships with others. Once generally doomed to failure, mainly because of the lack of physical proximity, long-distance relationships are now thriving like never before. But this is not the only example – all the relational aspects that underpin our civilisation have been transformed.

SUPREME INTELLIGENCE

Permanent and intrinsic hyperconnectivity in this new social context will also lead to a gradual transformation of the way our society works. This 'collective intelligence', which now allows instant and transparent information sharing, will lead to a better understanding of others' situations, of the world around us, and of the behaviours that drive us. And overall, to a better understanding of other cultures, and a more positive and empathetic relationship between peoples.

Political awareness will increase ten-fold, and this new paradigm will facilitate the emergence of new global communities around common interests. In some cases, these communities may reach the status of virtual states, no longer based on geographical limits, but on shared affinities, whatever their nature.

This does not mean that our civilisation has achieved total harmony. At least, not yet. Accessing these technologies – and the Ubernet in general – have not yet overcome the usual economic disparities. There is still a significant number of people in the world who can't access the Ubernet and are forced to evolve in a kind of disconnected underclass. The gap between the poor and the rich still exists, and the most pessimistic will probably tell you that it's unfortunately not going to disappear any time soon.

Moreover, despite the incredible technical progress made in recent years, human nature has not changed completely yet. We are currently seeing an increase in abusive behaviour (cyberbullying, blackmail), which we can undoubtedly attribute to the youth of this new technology. In addition, the compromises made about privacy to benefit from this new collective intelligence are still a stumbling block between those who are in favour of more transparency and sharing, and those who are still reluctant.

However, the promises of neural implants, combined with artificial intelligence and data collection, are tempting. We can dream of a day when – thanks to an advanced knowledge of quantum mechanisms – we will be able to control things by thought. Scientists at the beginning of the 21st century already envisaged this in theoretical form, but reality could quickly catch up with these old fantasies.

We are also not far from being able to digitise our consciousnesses, a giant step that would allow humans to exist outside the physical limits of the human body. Access to some form of immortality, through the digital transfer of the mind into a new android vehicle, will undoubtedly be the biggest revolution our species will experience. An ultimate form of transhumanism that would allow us, for example, to travel at the speed of light and conquer new exoplanets by downloading our consciousnesses into new bodies at the end of the journey. The Ubernet to conquer the Universe!

PROBABILITY

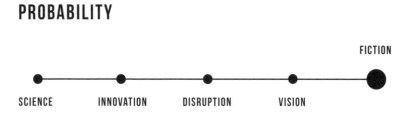

FICTION

SCIENCE   INNOVATION   DISRUPTION   VISION

This complete dematerialisation of the human being may still be science fiction today, but it would be dangerous not to take it seriously. The increasingly important role played by modern communication networks in our daily lives (social networks and others) has already led to deep changes in our society, with an impact many people don't fully understand.

If this revolution of a collective consciousness happens, it is likely to be far more disruptive than anything we've experienced before, and it would be dangerous not to properly prepare for it. With the miniaturisation of connected devices accelerating exponentially, the popularisation of nano-technology and neural implants is just around the corner. And with them, the promise of radical changes in the way our society works.

If we don't pay attention, there is a risk that a fringe of the population (if not almost all of it) will not be able to adapt quickly enough and in a relevant manner to the challenges that these new global and intricate networks will raise in the future. The best guarantee not to miss this civilisational shift, and to anticipate these drastic changes accurately, is to pave the way for this new paradigm of absolute connectivity as of today. If we wait for these new tools to be democratised before tackling the inevitable questions they'll raise, it'll probably already be too late.

FROM
DUSK TO DAWN

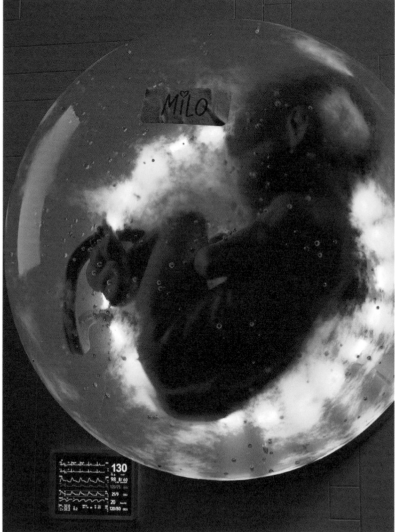

Olivia is on top of the world. Today, just as she was beginning to lose faith, the Medically Assisted Procreation (MAP) department at Singapore General Hospital finally called her regarding her in vitro gametogenesis (IVG) request. A few administrative steps later and her appointment is set for the 21st of January, 2051. Her dream of having a child is finally within reach.

The last few years hit her hard. When she was diagnosed with ovarian cancer in 2045 at age 38, she was forced to undergo uterine and ovarian ablation – just as she felt ready to take the plunge and have a child with her partner, Liam. A major blow. In February 2049, she contacted Singapore General Hospital in the hope of getting an appointment for IVG. After almost two years on the waiting list, her wish is finally coming true.

The day before her appointment, Olivia enjoys a night out with friends Levi and Aiden. These guys also went through this process to give birth to their baby girl, Lily. Olivia peppers them with questions about IVG and EUFI (extra-uterine foetal incubation). Although she relishes the chance to finally become a mother, Olivia is not at all reassured about the process.

Aiden eases her mind by explaining that the biopsy doesn't hurt at all. Lily was created from Aiden's skin stem cells that were transformed into eggs by IVG and then fertilised in-vitro by Levi's frozen sperm. She was then 'carried to term' in an artificial womb using EUFI. Levi explains that a midwife does regular follow-ups with the would-be parents at home. Other than that, they don't have to do much else besides alerting the midwife if anything seems wrong. On the day of the birth, the midwife ensures that the birth goes smoothly and records all the baby's data in a Life Pass that will follow the newborn for the rest of its life. Olivia and Liam will return home and start to enjoy their new lives as parents.

19 February 2051. It's D-Day for Olivia and Liam. A month after the collection of Olivia's stem cells, the couple is welcomed again at the hospital in a room right next to the lab where the biologists will create the egg cell. An augmented reality projection of what's happening under the biologist's microscope is shared on the glass window between the two rooms. The creation of the egg cell goes off without a hitch and it's placed in a small, fully sterilised rack alongside hundreds of others. It will develop in this giant incubator before being trans-ferred to its own a week later. The biologist explains to the couple that their future child will remain in the lab for two months. During this time, it will be connected to a EUFI incubator, and its development will be highly monitored.

At the beginning of April 2051, the midwife assisting the couple arrives at their home with the incubator. The foetus has developed well and will be able to continue its development outside the hospital. Liam and Olivia are delighted to be reunited with their future child and to be able to stay close to him as he develops. The midwife schedules another appointment with the couple mid-June to monitor the growth of the foetus and adjust EUFI settings.

20 December 2051. Olivia and Liam enjoy a quiet moment with their son in front of their holographic TV. Little Milo is now one month old and thriving. The new parents get a notification from the Life Pass app on their phones to say that an appointment was automatically made with their family doctor for his first TB vaccination. It's 1 pm and the news starts with the headlines: the CO_2 level in the air in Beijing has dropped by more than 50% (an all-time record); the remastered film Invictus is released, celebrating sixty years since the end of Apartheid and in which Morgan Freeman is posthumously played by a digital actor. As Milo falls asleep in Liam's arms, one headline in particular catches the couple's attention. Cornell University's Reproductive Endocrinology Laboratory has just taken reproductive research one step further. The first child with eight parents was born in perfect health – each parent from a different part of the world. With its multiple origins, this child's facial characteristics may just be the future face of humanity.

IVG AND EUFI: TO INFINITY AND BEYOND

By 2051, virtually anyone with a skin would have the ability to produce eggs or sperm. Procreation has always been a somewhat taboo subject that we don't like to medicalise. Just take a look at the vitriolic reactions people had to the advent of MAP in 1978. And yet, forty years later, we've had more than 8 million births thanks to this new discovery. Since then, research has continued to move forward and improve to allow everyone to procreate – re-gardless of gender or sexuality. In 2051, if you want to have a child, you can do so naturally but almost half the population will first make an appointment with a MAP hospital service. Research improvements mean that infertility no longer exists. Procreation is now possible for absolutely everyone.

Since 2042, thanks to IVG coupled with EUFI, a same-sex couple can conceive a child without the intervention of a third party and become biological parents. Skin cells from one half of a couple will be transformed into pluripotent stem cells. These cells will then be reprogrammed into male or female gametes (reproductive cells), depending on the missing type of gamete. Women who no longer have ovaries or whose oogenesis (the process of an egg cell becoming an ovum) isn't functioning properly can also have their stem cells transformed into gametes.

IVG opens up a multitude of possibilities and allows everyone to fulfil their desire for parenthood. This has been taken a step further since the end of 2049 when people expressed the wish to procreate on their own. IVG allows them to create both male and female gametes from their skin.

This hasn't been approved by the international bioethics committee, as the very limited genetic mixing could be risky for the child. However, IVG has paved the way for so-called multiplex parents. Last year, in early 2050, four polyamorous people were able to give birth to the first four-parent child. The four parents-to-be – in this case, two men and two women – were separated into two pairs whose gametes were harvested to form two embryos through in vitro fertilisation. One embryo was then reprogrammed into a male gamete, the other into a female gamete. IVF of the egg by the sperm, both artificial, allowed the creation of the final egg cell. The development of the foetus and the birth of the child went well, but its growth is still monitored to prevent all possible problems.

Creating gametes and egg cells to grow an embryo has been made possible. But how can we ensure the development of this embryo without an intra-uterine environment? Quite simply – by creating an artificial environment. This has taken decades of research, as the placenta is an extremely complex organ to re-create, and amniotic fluid has to respect extremely specific conditions in terms of composition and temperature.

But with the number of very premature babies increasing every year, researchers around the world have been looking into the matter. In 2038, after numerous tests, the first very premature baby of 19 weeks was able to grow to term. Thanks to a state-of-the-art incubator that identically reproduces the intra-uterine environment, this replicates the exchanges between the foetus and the body of the mother through the placenta: the EUFI. Since the child had no side effects and all its vital signs had stabilised once in the incubator, this process was taken a step further. At the end of 2040, the first child developed entirely with EUFI was born, opening a whole new field of possibilities. However, this technology remained exclusively for medical purposes until 2046. EUFIs are now available to everyone – subject to approval and a certain waiting period.

The availability of EUFI is a new milestone in access to reproduction, whether for same-sex couples, for women who don't or no longer have a uterus, or for women who wish to have a child without having to carry it. This marked the end of the seemingly unbreakable link between sex and procreation. It has also brought great societal upheaval since reproduction no longer relies mainly on women.

THE LAB: THE FUTURE OF PROCREATION

And why not? Since 1978, IVF has proved its worth. In addition to the millions of babies born through this method, women as old as 70 have been able to give birth. In some countries, same-sex couples can also become mothers with the help of a donor. In 2016, IVF opened the door to multi-parenting with the birth of the first three-parent child, which enabled a woman with the hereditary Leigh syndrome to give birth to a healthy child.

IVG is a science to take seriously. In Japan, a group of researchers in agriculture, medical and embryo science conducted a promising experiment on a mouse in 2016. Its eggs were artificially created from skin cells. As a result, she gave birth to 26 pups.

On a human level, research is still in its infancy. Japanese researchers tried to create an egg cell from blood cells. Unfortunately, the gamete was too immature to be fertilised, but scientists still managed to create the egg. We're not at the stage where we can start thinking about multiplex parents yet, but the beginnings are promising. We know the process works, but it will take a few more years of research to be applied to humans.

As far as ectogenesis is concerned – the development of an embryo or foetus in an artificial womb – opinions are divided. Magdalena Zernicka-Goetz and her colleagues at the University of Cambridge have succeeded in developing an embryo in-vitro for 14 days. On the other hand, maternity wards are able to take care of very premature babies at the 22nd week of pregnancy (i.e. five and a half months).

The question is: will we be able to close this 5-month gap and ensure artificial foetus development for a whole nine months? Scientists are hard at work to make this happen. Yoshinori Kuwabara, professor at Juntendo University in Tokyo, has developed a technique called EUFI. Using EUFI, he can ensure the proper development of goat foetuses for three weeks using catheters connected to the umbilical cord and incubators containing artificial amniotic fluid.

PROBABILITY

FICTION

SCIENCE INNOVATION DISRUPTION VISION

Science is progressing fast, and it looks as if IVG and artificial gestation may arrive sooner than we think. It's going to open a hornet's nest of legal questions and ethical dilemmas that we should start thinking about already. How many parents should we allow to be involved? Is it going to be accessible to everyone without any stipulations? Should IVG be a choice for all, or an absolute last resort? And how will IVG influence the development of future generations – halting adoptions, creating so-called 'designer babies' and causing a further population explosion?

It's one vision of the future which could mean immense joy for some, and massive conundrums for rule-makers around the world.

GOODBYE
MORTALS

Marianne is packing eagerly for her trip. Trousers, shorts, t-shirts, jeans, a light jacket for cool evenings. Two bikinis, a slinky cocktail dress, her favourite heels. Abseiling kit, check. Cycling kit, check. Lastly and most importantly, her Rejuva supplements and nanomedicine – packed in a big temperature-controlled case for her 3-week trip. Marianne is travelling to her alma mater, the University of California, to attend an open day for a degree in mind-transfer technology. Since transferring her own mind onto the cloud, she's become obsessed with the field. Before hailing a flight taxi, she quickly checks herself in her smart mirror. Her blonde hair is freshly tinted and bouncy thanks to treating herself to some 4D-printed hair follicles. Her eyes are bright blue and clear, her face only lightly wrinkled, thanks to religiously wearing her sun shield suit outdoors. She spritzes some pheromone perfume on her ample cleavage, dabs algae gloss on her plump lips and flashes herself a toothy smile. Not bad for an 85-year-old, she thinks, as she smoothes down her mini pencil skirt. She's always been a bit self-conscious about her missing canine teeth but the new wonder gel that regrows missing teeth has worked wonders. They even match her other pearly whites.

Tonight, she's meeting up with Frank – a 76-year-old she met on the virtual reality dating app, Octo (cheekily named after the old-fashioned word 'octogenarian'). They've had lots of hook-ups in virtual bars and hotels (shh...don't tell the grandkids) but this will be their first face-to-face rendezvous. After her night of fun in LA, she'll meet up with some friends and go rock climbing in Joshua Tree National Park for a few days. This will be such a bucket-list moment for her – the first time she's fit and able enough to conquer those boulders the way she used to as a 20-year-old.

When Marianne was in her 60s, she developed a rare motor neuron disease that would normally have her confined to a wheelchair. She was fortunate enough to take part in a gene therapy trial that halted her paralysis, yet she still ended up walking with difficulty and living with an assistance robot for many years. Doctors kept telling her what a miracle she was – that she was still alive, could still walk and talk, and enjoy her job as data detective.

But for the fit and active Marianne, it wasn't enough. She was over the moon when a new stem cell therapy – mimicking the rejuvenation of a rare seahorse – became available. After 10 years and countless procedures, her defective genes have been removed and her muscle strength has ever so slowly increased. With the help of precision nanomedicine, tailor-made nutrition and muscle shock therapy, she has gone from managing one sit-up a day to running 10km each day. She looks and feels amazing, and with 20-30 years ahead of her, thinks it's the perfect time for a new career path. Why not? Life begins at 90!

CONQUERING THE BIGGEST KILLER OF ALL: AGEING

'There's no chance for us / It's all decided for us...Who wants to live forever...' – so the Queen song goes. Death is inevitable, right? Not in 2050. Scientists are confident that death will be a choice in as little as three decades' time. Immortal? No, come on. We're not vampires or superheroes or jellyfish (yet). We'll be what the transhumanist movement calls 'amortal' – where the radical lengthening of your lifespan in good health will be a conscious choice.

No-one is going to force you to live for 200 years. Like everything in life, it's up to you. You can choose to smoke and knowingly shorten your life. If you start to find life boring, you can choose to let it run its course. Or you might be the sort of person who wishes for 48 hours in every day. Someone who needs 500 years to cram in all the living you want to do. For you, there is only good news. Either way, when you die, it will be a death in your time frame, not a death suffered. Long before 2050, you'll start having regular checks on your ageing mechanism risk factors by your doctor. Depending on your stats, you'll be prescribed various preventative treatments, from nanomedicine to automatically correct your blood pressure, to gene therapy slowing down greying hair or osteoporosis. There will be treatments for your senescent (old, inactive) cells or your autophagy (the process whereby your body gets rid of old, damaged proteins). Everything will be constantly monitored and adjusted to help you live happily and agelessly.

The hallmarks of ageing – dementia, Alzheimer's, diminishing eyesight, cancer, heart disease – will now be managed effortlessly like you would arthritis or diabetes. 'Brain pacemakers' (brain stimulation implants) will be as ubiquitous as cardiac pacemakers. Currently they treat epilepsy and movement disorders like Parkinson's; in 2051 you could get deep brain stimulation for depression, OCD, memory impairments, back pain and much more. Implants will come with AI technology that can observe brain function and make adjustments on the spot. Rejuvenation therapies will be widely available – not just to the wealthy elite – as open-source technology is shared freely to anyone to adapt. You'll no longer have to live with DNA you inherited from your parents. Genome editing will be so far advanced that you could prune away at any undesirable DNA cells or rewrite new ones. Gene splicing will add new genes into your body – for example, photoreceptor cells to restore vision in blind people. If an organ gives out, you will no longer have to wait months for a suitable donor – one could be 3D-printed for you. Suffering from dodgy dental work from decades back? No more scaring the grandkids with your false teeth. You can simply grow back the teeth you lost. Science fiction? Definitely not. Welcome to the age of amortality.

SOLVING THE PROBLEM OF AGEING

Let's face it – ageing is going to be solved very soon and likely by Silicon Valley. Longevity has become an obsession for tech titans who are throwing mountains of cash at the problem of lifespan extension. Jeff Bezos, Peter Thiel, Sergey Brin, Larry Ellison – they are all either funding longevity research or experimenting with anti-ageing interventions themselves (or both). If any-thing can be hacked, why not the process of ageing? It's happening already and we'll see exponential growth in this area over the next few decades.

But first, we must figure out exactly what causes ageing. The first topic of study: the senescent or so-called 'zombie cells' in our bodies. These cells stick around forever, not really doing much, along with your other healthy cells. However, researchers now see that senescence plays a very important role.

As we age, these damaged cells start to accumulate and cause havoc – changing your metabolism and stem cell function, kicking off the ageing process and accelerating conditions associated with it such as Alzheimer's disease. If we can destroy them, we can live longer.

In 2016, this was proven in an animal trial performed by Jan van Deursen, a molecular biologist at the Mayo Clinic. The study showed that if you purge senescent cells in a mouse born on the same day, from the same litter, and raised in the same conditions as its siblings, it will make him look younger and will delay the onset of mice-related ageing like cataracts and a bent spine.

Pharmaceutical companies and investors are working hard to find the compounds that can target senescent cells and are currently doing human trials with 'senolytic' drugs. One such company that is committed to producing anti-ageing medicine very soon is Unity Biotechnology (of which Van Deursen is a founder). This San Francisco start-up has attracted over $200 million from investors that include PayPal co-founder Peter Thiel and Amazon CEO Jeff Bezos. Unity's tests currently focus on delivering localised therapy in ophthalmologic and neurologic diseases. Its stance is that when you view ageing through the lens of specific diseases, you will make someone more functional and therefore able to live longer.

Others feel that larger trials should be done on 'geroprotectors' – the drugs that seem to slow down ageing altogether. The star here is metformin, the low-cost diabetes drug that may have already saved more people from cancer deaths than any drug in history. Comparing the results of diabetics taking metformin over twenty years versus those on other diabetes drugs, the results were astounding. Metformin takers were all-round healthier: they lived longer and had fewer cardiovascular events. They seem to suffer less from dementia and Alzheimer's. But the biggest bonus was that they were 25-40% less likely to contract cancer than diabetics on other popular medications. Even when they did, they outlived those on other medication.

No wonder then, that Silicon Valley execs are popping metformin like candy – side effects be damned. Many scientists are pushing for the wonder drug to be approved for widespread use sooner rather than later. There is even data that shows it can significantly reduce the risk of death from Covid-19 in women.

The main issue is that the FDA doesn't recognise ageing as a disease. That's why the TAME Trial – Targeting Aging with Metformin – is so exciting. While focusing on the three main age-related chronic diseases (heart disease, cancer, and dementia), if the trial is successful it will be the first time that ageing is made an indication for treatment. We'll only have those results by around 2027.

Looking much further ahead is the game plan of another longevity frontrunner, Calico Labs. Funded with billions of dollars by Google parent company Alphabet, Calico is on a mission to deeply understand the process of ageing with a very long-term view. For example, they are accurately mapping the genome of the naked mole-rat, which is able to live ten times longer than a mouse. If we can understand how the mole rat's genes are organised for such a long life, maybe we can do the same for humans.

Analysing species with negligible senescence (biological ageing) is one of the most fascinating areas of research. Plants and animals in this category don't show any decline in functional or reproductive capability, or rising death rates with age. On the almost mind-boggling side of the spectrum is the Great Basin Bristlecone pine that can live over 5000 years. Under the sea, the ocean quahog clam lives well over 500 years and Greenland sharks often celebrate their 400-year birthday. As for mammals, naked mole-rats have the longest lifespans relative to body size of any known, non-volant mammalian species.

Then there is of course the famous immortal jellyfish that can turn back time and defy death. Like many lobsters, sponges, corals and hydras, the Turritopsis dohrnii doesn't degrade as it gets older, its fertility remains constant and it can regenerate after an injury and even reproduce asexually. Put an immortal jellyfish under stress and it won't start comfort eating like us humans. If sick or old, attacked or under environmental stress, the jellyfish reverts to its polyp stage. (Basically, it's like a butterfly moving back into a cocoon.) The jelly-fish does this through a cell development process of transdifferentiation – changing the state of cells and transforming them into new types of cells.

How does this anti-ageing actually happen? It's all down to the work of telomerase. This enzyme repairs long, repeating sections of DNA sequences at the ends of chromosomes, called telomeres. If we can figure out how our own human telomeres can be manipulated, it will, in the words of scientist Shin Kubota, be 'mankind's most wonderful dream'. Mice bred with telomere lengthening have been proven to live longer, and molecular biologist Maria Blasco has developed a telomerase-based gene therapy for the treatment of different pathologies related to the shortening of telomeres – in other words, ageing.

Founder of the Singularity University, Ray Kurzweil, predicts that we'll hit 'longevity escape velocity' as soon as 2030. This is going to be a profound revolution – the point at which, for every year that you're alive, science will be able to extend your life for more than a year. Biotechnology will take over from medicine and we'll be able to seek and destroy cancerous cells and repair damaged organs. By the 2040s, we will be living agelessly and rejuvenation therapies will be affordable to everyone.

PROBABILITY

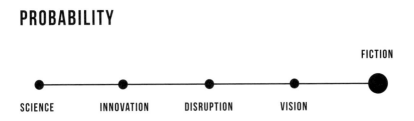

FICTION

SCIENCE   INNOVATION   DISRUPTION   VISION

Centuries ago, death was simple. You died from an infected wound, childbirth or the flu. Today, thanks to much better nutrition, vaccines and cutting-edge medicine, life expectancy has doubled in wealthy nations. By 2050, the global population age 65 and older is projected to nearly triple, to 1.5 billion.

Ageing is the biggest pandemic we've ever had and the number one killer of the human race. Yet, there will always be people who'll say just because we can extend life, does it mean we should? Is it ethical? In their book, *La Muerte de la Muerte* (*The Death of Death*), renowned futurists José Cordeiro and David Wood argue that if it's moral to cure cancer, Parkinson's and Alzheimer's disease, curing ageing is the most ethical thing we could do. If anything, it's something that should've happened much sooner already. If we've known for almost seventy years that cancer cells are immortal, why is there no cure for cancer yet?

The choice to live disease-free for longer is a good thing. Whatever we do, we can't expect the world of 2050 to be anything like the world we know today. Adding another 20-40 years to our life expectancy will take a mind shift – a move away from the notion that the life of an elderly person is somehow less important or that one must disappear into non-existence after a certain age. There is nothing wrong with continuing to work and contribute to society (and the economy) well into your 80s and 90s. The cut-off point should be a deeply personal choice for each person – and certainly not determined by preventable diseases.

CONCLUSION

BELIEVE
IN YOUR OWN RELIGION

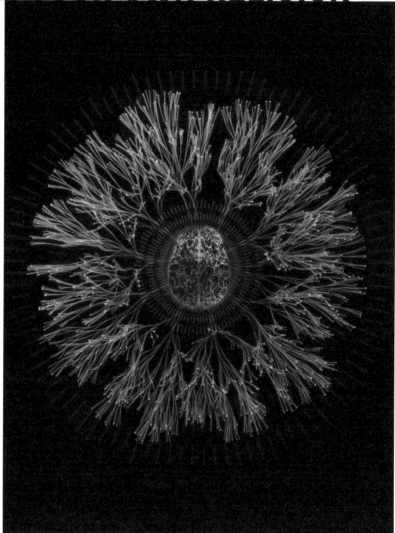

By
Brice Le Blévennec

Have you heard of the Ultimate Supercomputer? It boasts mind-blowing storage capacity, excellent memory, top energy efficiency and super-fast processing speed. It acts as a powerful interface between humans and the physical world, transforming inputs into outputs, encoding and storing information. It constantly processes and reprocesses information – even when it's on sleep mode. It has a highly sophisticated way of recalling memories and experiences that happened as many as 60 or 70 years ago. Even occurrences that happened during factory assembly. It has an inimitable way of communicating with other computers – millions of them around the world. The best thing about this supercomputer? It's already living in your home.

You've guessed it: I'm talking about the human brain. Compared to a computer, the brain is ten times more energy efficient than a computer. Our brains only use roughly 10 watts of power, compared to the 100 watts of a typical computer. With an estimated 200 billion neurons, the brain's storage capacity is thought to be 1 petabyte – the equivalent of 1000 one terrabyte SSDs. While you might argue that you can't compare computer bytes with neurons, remember that every single neuron is linked to others via hundreds of trillions of tiny contacts called synapses, along which electrical impulses travel. In the cerebral cortex there are more than 125 trillion synapses. It has the unique ability to combine one neuron with another to help create and store memories. We might not have the accessing memory of computers but our ability to relate and string memories together is unsurpassed.

Why am I telling you about brains? Because they are not computers, they are better. A brain's complexity is almost beyond belief – an incredible organ filled with a rich tapestry of memories and intricate connections. Brains can think thoughts, churn out ideas and solve problems. Supernatural abilities are not just the remit of magicians, mind readers and fortune tellers. Some people with autism have abilities beyond the range normally experienced by 'neurotypicals'. Some, like activist Greta Thunberg, call these superpowers – unusual talents, skills, qualities and advantages.

The organisation Embrace Autism has a long list of sensory, cognitive and behavioural strengths. For example, autistic children can see a greater intensity of colour. Visual hypersensitivy means that autistic people can see from six metres what a non-autistic can see at two metres. They have acute hearing and heightened pitch detection. They very often have high prevalence of synaesthesia (where multiple senses are perceived simultaneously, e.g. see smells, taste sounds). Then there is the encyclopaedic knowledge, savant-like memory and profound abilities in music, art, mathematics or all things mechanical.

There's also the small percentage of people born with 'savant syndrome' – an extraordinary condition where someone with a serious mental impairment displays spectacular abilities. Megasavant Kim Peek could read by scanning the left page with his left eye, while reading the right page with his right eye. Peek, who only walked at age four and couldn't button up his shirt, could accurately recall the contents of at least 12,000 books. Then there's architectural artist Stephen Wiltshire who, after just one short helicopter ride over a city, would create a detailed drawing of it. In 2005, Wiltshire produced his longest ever panoramic memory drawing of Tokyo on a canvas of ten metres long.

Savants usually encountered brain damage, generally in the left side of the brain. To compensate, undamaged parts of the brain and lower-level memory capacities are engaged. A kind of rewiring happens and releases dormant capacity from the newly wired area. This process is described by the world's foremost authority on savants, Dr. Darold Treffert, as 'the 3 Rs': recruitment, rewiring, release – when exceptional abilities that were previously dormant rise to the surface. It has been known to happen later in life too, called 'acquired savant syndrome'. Following a head injury, dementia, stroke or other form of brain damage, prodigious skills in art and music suddenly appear. There are many examples that Dr Treffert is keeping track of, such as the orthopaedic surgeon from New York who discovered a passion for playing the piano after being struck by lightning, or the man who began to paint and write poetry after suffering a stroke.

This ability to know things that you never learnt is fascinating to us. Dr Treffert believes that all of us have this latent *Rain Man*-like capacity inside us. He's not the only one. The Australian neuroscientist Allan Snyder has a vision of a 'creativity cap' – a method of transcranial magnetic simulation that can artificially replicate savant syndrome. With some mild brain stimulation, it suppresses brain activity in the left anterior temporal lobe while exciting activity in the right anterior temporal lobe. The studies never got anywhere but, who knows, maybe in 30 years' time we'll all have access to brain-boosting caps?

I believe that we are gifted with these powerful processors inside our bodies for a reason. Our brains perceive so much more than what our conscious mind sees, hears, smells and feels. Our environments matter even before we're born. Firstly there is our DNA – the traits we inherit from our parents but also the instruction manual on how to grow cells, find food, procreate, etc. This is where the relatively new field of epigenetics is so interesting.

In the 20th century, scientists started to realise that it's not only our DNA sequences telling genes what to do. Some instructions were coming from our epigenome – a collection of other chemical markers and signals that influence our DNA. We've seen studies of mice where fear memories were inherited across two generations. Researchers trained male mice to associate a certain smell with an electrical shock, so that after a while they would fear the odour itself. Would you believe it, they then saw that the smell also scared the next two generations of mice. In another study, researchers found that the grandchildren of women who were pregnant during the Dutch famine of 1944 to 1945 were more likely to be fat as newborns. Coincidence? Perhaps. But we're going to discover much more about this epigenetic inheritance over the next decades. We may well learn that someone can inherit experiences.

DNA aside, our first experiences start in our mothers' wombs as we soak up all of her habits. We're born as empty vessels and start digesting our own feelings and observations. Don't think for a moment you only start having memories from age five. Everything that happens to you from the moment you open your eyes is carefully filed away in your conscious and subconscious mind. As you're bombarded by thousands of pieces of micro-information, your brain does its astounding processing thing in the background, constantly monitoring your ambient surroundings.

DREAM SOME SENSE INTO LIFE

At night, as you dream, you relive the emotions of the previous 24 hours. Your body goes through the filing cabinet of the day and decides which of these learnings, likes, dislikes and experiences to keep, and what to discard.

As your brain synapses make connections, every piece you store is related to a smell or sound or feeling, compartmentalised to what your subconscious deems important. What you dream might not be real but the emotions attached to these experiences certainly are. In essence, dreams help us to strip the emotion from an experience by creating a memory of it.

Did you know that you don't see colour or faces in your dreams? It's when you save them as memories that you 'paint over' colour and assign meaning. We are all quite focused on what happens in the foreground of our lives but I believe it's what happens in the background that is really powerful. Says researcher of hypnagogia (the brief time between waking and sleep), Valdas Noreika: 'When we enter sleep, the brain steadily dismantles the models and concepts we use to interpret the world, leading to moments of experience unconstrained by our usual mental filters.'

These unconstrained experiences are the heart of many of the seemingly inexplicable things that happen to us: out-of-body experiences, déjà vu, telepathy, daydreams, intuition... This is the 'whispering in your ear' our body does when we draw on our memories for guidance. It informs our humour, how we fall in love, what gives us anxiety, or why a song or a painting will make you cry or feel angry or turned on.

INTERFACE US

Nature is superb at building transducers, according to senior research psychologist Robert Epstein. In his essay about the brain as a type of transducer – converting one signal to another signal, from one medium to another – he notes that our bodies are encased in transducers from head to toe, especially our sense organs (eyes, ears, nose, tongue, and skin) which transduce things like air pressure waves, textures, pressure, and temperature into distinctive patterns of electrical and chemical activity in the brain.

Epstein floats the interesting notion that the human brain is a transducer that connects us to some kind of 'Vast Intelligence' or 'Operating System'. How else do we explain things like blindsight (the ability of some blind people to be aware of objects in their environment that they cannot consciously see) and mindsight (when congenitally blind people have near-death experiences in which they are able to see normally)? What happened with the Australian woman who, after surgery, woke up with an Irish accent? Called 'foreign accent syndrome', could this super rare condition be down to a transduction error?

How we connect with other human beings is what gives meaning to the world. This is our religion. I see the way we connect with others as sets of overlapping circles. What's inside each of our circles is what you choose to fill your life with: feeding your mind with thought-provoking content (not just TikTok), hanging out with people that challenge your thinking, cross-pollinating your being with interesting concepts. Pruning toxic situations from your life and using the power of technology to fertilise ideation.

We are more connected in our thinking than I think people realise. We all breathe the same air, drink the same water, feel the same sun on our skin. We're part of the noosphere – the realm of human minds interconnected through communication. The noosphere sees human cognition as a part of the environment of Earth. As the prophet of cosmic hope, Pierre Teilhard de Chardin, explained it: first there was the geosphere (the 'rock' of the world and the gas around it). Then there was the biosphere – the ever-changing biochemical make-up of life. And finally there was the noosphere – when one of these biological creatures developed the mental and practical power to change the world in ways that no other creature could. De Chardin believed that human reason and scientific thought have created the next evolutionary geological layer.

2050: EVOLVING THROUGH CREATIVITY AND IMAGINATION

'*The future belongs to those who give*
the next generation reason for hope'.

Pierre Teilhard de Chardin

My vision for 2050 is a world where people are finely tuned to their unconscious minds. A world where we can bear witness to the extraordinary resilience of the human spirit and are shown that, even in the most desperate circumstances, there can be hope. I hope for a world that we will take our destiny in our hands and continue to use the incredible creativity and imagination that we humans are capable of to evolve our civilisation further. As de Chardin declared: 'Each act of love, no matter how small or hidden, moves all of reality closer to unity and connection.'

By
Brice Le Blévennec

ACKNOWLEDGEMENTS

This book is a valiant team effort from the whole Emakination – the entire global staff of Emakina, the User Agency. More than fifty Emakinians from twenty countries (in which we are present) contributed their ideas and visions to the chapters in this book. I had a wonderful time discovering the future through their eyes and was often impressed by the foresight of many of these predictions. What was even more fascinating was that some of these prophecies came from the world's furthest corners, from Sweden to South Africa to Serbia. Every single theme suggested was worth pondering, and we were careful not to leave any of them out.

I would like to personally thank the team that put their heart and soul into turning these ideas into thirty exciting chapters.

Starting with **JOHANNIE VAN AS**, whose vivid imagination and ability to delve deeply into a subject she has only just discovered often astounded me. She has forced me to put my visionary talent into perspective and I have come out of this creative relationship much more humble than when I started. We created half of the articles together but I have to say that if I lit the spark, it was she who burned brightly.

She introduced us to her equally talented partner **PAULA FITZHENRY**, to whom we owe some of the most exhilarating fictional pieces in the book, plus a masterful copy edit of text from the broken English of Europeans into smooth native English.

Equally impressive is **SARAH CLAEYS**, whose literary style is delectable with a sense of detail that makes the storytelling so realistic it's completely believable. I devoured her writing, unable to offer any constructive criticism (even though I was supposed to read them again to check that they fit the vision). Sarah has been in charge of Emakina's digital content experience for many years. Since this project, with me in the client position, I now completely understand why all our clients adore her.

MANON DUBREUIL impressed all of us with her writing talent, which was a real revelation for her and the whole team. And to think that she didn't know she was capable of creating fiction in both French and in English (which is rare for a french person)... I am sure that this exercise has ignited a passion for writing which will not be extinguished any time soon.

I would also like to thank my friends **JEAN-CHRISTOPHE DETRAIN** (known as Faskil) and **CEDRIC GODART** who supported us with their professionalism at the start of this book. Without them, we probably never would have had the courage to take it on, overwhelmed by the enormity of the task at hand.

Finally, how can I not salute the essential contribution of **IVA FILIPOVIC**, who went through the entire web with a fine-tooth comb and watched all of Youtube (and TikTok) to prove that our crazy ideas were often already budding realities. She also had the strength to challenge me (and, above all, to be right) on many occasions. This saved our readers from a lot of really crazy ideas... which I'm saving for a future novel one day.

A book is not just literature, it is also a real project, supported by a team of Jira ticket killers. Without my funny ladies, **CHLOÉ MARCHANT**, **ÉLÉONORE CALICIS** and **MANON DUBREUIL** (who has several strings to her bow), we never would've met our tight publishing deadlines for this book – our gift for Emakina's 20th anniversary. It probably would have been postponed until Emakina's 30th anniversary! It is also an illustrated book made up of original creations by real artists whose day jobs might be Emakina designers but who could make a fortune on the art market! Actually, on second thought, I take that back. I would much rather they stay with us and continue to impress our clients with their talent.

This is finally a publishing project and I must thank **MICHELLE POSKIN**, whose patience, kindness, competence, sound advice and a small dose of harassment have been instrumental in the success of this book.

CONTRIBUTIONS. @Emakina.AE: Guillaume Loiseau, Sales force marketing cloud lead / Huzefa Tarwala, Project manager – Emakina.AT: Irina Voehr, Marketing and communication manager – Emakina.BE: Aline Durand, Head of insights / Éléonore Calicis, Marketing and communication project manager / Estelle Le Nestour, Designer / Iva Filipović, Experience design researcher / Johannie van As, Copywriter / Lore Donné, Front-end developer / Maarten De Neve, Marketing automation expert / Manon Erb, Marketing assistant / Maxime Honhon, Project manager / Pierre Boulanger, Growth hacker / Rani Nasr, Account manager / Sarah CR Claeys, Copy & Content strategist / Stefan De Haes, Project manager / Thomas Van Roy, Social media manager / Vicky De Mesmaecker, Application design strategist / Yoeri Conickx, Account manager – Emakina.FR: Alexandre Jardin, Software architect / Alexis Mons, Head of legal and GDPR / Bertrand Duperrin, Head of people and business delivery / Jérôme Depaifve, Scrum lead / Nicolas Borgis, Managing director / Romain Dehaudt, Head of revenue & operations – Emakina.HR: Valentina Zanetti, Project manager – Emakina.NL: Alex Veremij, Performance specialist / Bas van den Biggelaar, Marketing specialist / Bugra Merhametli, Project manager / Daan van den Berg, Account manager / Domiziana Luzii, Visual designer / Eef Poetsema, Project manager / Oguz Boz, UX intern – Emakina.PL: Marcin Średziński, Back-end developer – Emakina.RS: Aleksandra Miletić, Front-end developer / Andrija Cvejić, Software and machine learning engineer / Djordje Stefanović, Sales force developer / Dragan Špančić, Designer / Luka Bjelica, Front-end developer / Maja Mišović, Marketing manager / Spasoje Malbaški, QA engineer – Emakina.TR: Onur Özgüzel, NET developer / Uluc Guralp, Financial controller – Emakina Group: Alexandros Papanastasiou, Integration director / Brice Le Blévennec, Chief visionary officer / Chloé Marchant, Marketing and communication director / Luc Malcorps, Director of Media Relations / Manon Dubreuil, Content manager – The Reference: Frank De Graeve, Functional analyst

ILLUSTRATIONS. @Emakina.BE: Audrey Zaludkowski, Designer / Christophe Deaconescu, RIA developer / Dimitri Vankerkoven, Designer / Eloïne Philippe, Designer / Estelle Le Nestour, Designer / Georgios Leontiou, Designer / Kenan Murat, Designer / Luca Petolillo, Designer / Marc Dalemans, Designer / Patrick Jones, Motion designer / Quentin Baes, Art director – Emakina.RS: Sara Babić, Designer – Emakina.SE: Johan Bränström, Creative – Emakina.ZA: Erin Kemper, Art director

© Éditions Racine, 2021 / Éditions Racine, Tour & Taxis - Entrepôt Royal Avenue du Port, 86C / bte 104A / B-1000 Bruxelles
D.2021.6582.32 / Legal deposit: December 2021
ISBN 978-2-39025-187-3